MW01043391

Computing Strategies for Reengineering Your Organization

Other Prima Computer Books

Available Now

WINDOWS Magazine Presents: Access from the Ground Up
Advanced PageMaker 4.0 for Windows
DESQview: Everything You Need to Know
DOS 6: Everything You Need to Know
WINDOWS Magazine Presents: Encyclopedia for Windows
Excel 4 for Windows: Everything You Need to Know
Excel 4 for Windows: The Visual Learning Guide
WINDOWS Magazine Presents: Freelance Graphics for Windows: The Art of Presentation
Harvard Graphics for Windows: The Art of Presentation
Improv for Windows Revealed! (with 3½" disk)
LotusWorks 3: Everything You Need to Know
Microsoft Works for Windows By Example
NetWare 3.x: A Do-It-Yourself Guide
Novell NetWare Lite: Simplified Network Solutions
PageMaker 4.0 for Windows: Everything You Need to Know
PageMaker 4.2 for the Mac: Everything You Need to Know
PageMaker 5 for Windows: Everything You Need to Know
PC DOS 6.1: Everything You Need to Know
WINDOWS Magazine Presents: The Power of Windows and DOS Together, 2nd Edition
Quattro Pro 4: Everything You Need to Know
QuickTime: Making Movies with Your Macintosh
Smalltalk Programming for Windows (with 3½" disk)
Superbase Revealed!
SuperPaint 3: Everything You Need to Know
Think THINK C (with two 3½" disks)
Windows 3.1: The Visual Learning Guide
Windows for Teens
Word for Windows 2: The Visual Learning Guide
Word for Windows 2 Desktop Publishing By Example
Word for Windows 6: The Visual Learning Guide
WordPerfect 5.1 for Windows Desktop Publishing By Example
WordPerfect 6 for DOS: The Visual Learning Guide
WordPerfect 6 for Windows: The Visual Learning Guide

How to Order:

Individual orders and quantity discounts are available from the publisher, Prima Publishing, P.O. Box 1260BK, Rocklin, CA 95677-1260; fax (916) 786-0488. If you are seeking a discount, include information on your letterhead concerning the intended use of the books and the number of books you wish to purchase.

Computing Strategies for Reengineering Your Organization

Currid & Company

Prima Publishing
P.O. Box 1260BK
Rocklin, CA 95677-1260

Prima Computer Books is an imprint of Prima Publishing, Rocklin, California 95677

Executive Editor: Roger Stewart
Managing Editor: Neweleen A. Trebnik
Project Editor: Stefan Grünwedel
Copyeditor: Bob Campbell
Production: Susan Glinert, BookMakers
Indexer: Brown Editorial Service
Book Designer: Susan Glinert, BookMakers
Cover Designer: Page Design, Inc.

Library of Congress Card Number: 93-085965
Printed in the United States of America
94 95 96 97 RRD 10 9 8 7 6 5 4 3 2 1

Dedications

To Ray, Tray, Justin, and Nanny. Thanks again.

—Cheryl Currid

To Brian and Nicole, the people who make me look forward to "time off." Also to Irene and John Chmiel—Mom and Dad—who don't understand computers but believe in what I do anyway.

—Linda Musthaler

To my wife, Barbara, whose love and devotion serve as a constant source of inspiration.

—Frank Ricotta

Contents at a Glance

Table of Contents

Acknowledgments

Every book turns into a team effort—and this one is no exception. Our research and interviews for this book took us to some very interesting places and allowed us to talk to many interesting people.

First, we'd like to thank the efforts of our editors and publishers at Prima Publishing for undertaking this project. We believe this book will help unravel the confusion surrounding corporate computing options, and we appreciate their sharing the vision and bringing it to market.

Next, we'd like to acknowledge the extra effort of Currid & Company's immediate and extended family. Kent Drummond played a helpful part in gathering early research on specific corporate computing issues. Diane Bolin spent many a late night reading (and re-reading) early forms of the manuscript. Also, Diane's research abilities were a great help when we needed to dig out just a "few" more facts. Tony Cross and Josh Poniard both contributed significantly to several chapters—and to the sanity of the authors. Dianne Davison contributed her wisdom and cool, calm, and collected demeanor during all the little crises that took place.

We'd also like to acknowledge other contributions. Rather than make a lengthy list of our sources, we'd like simply to say thank-you to the many corporate business and information services managers who contributed their time and war stories to this effort. On several occasions, we kept them up late nights telling us about their experiences, opinions, and feelings. For those who wished to remain anonymous, we've honored your wishes. Finally, we'd like to thank our families for their patience and understanding during this project. Book projects simply aren't 9-to-5 jobs.

Introduction

This book can change your life, your company, and your feelings about computers in business. That's its purpose. It lets you in on the latest thinking about what you can and should do with your computers.

In short, this book provides you with a road map to using the new information technology as you reengineer your business processes. Beyond that, the rest is up to you.

Computers have left an indelible mark on businesses around the globe. They are here to stay, and they change everything—we hope, for the better. The more effectively you learn how to use computing technology, the more you can change things. Hundreds of companies have found that computers hold the keys to their competitive advantage, better decision making, superior customer service, and faster time to market. Other companies have spent big bucks buying computers but have failed to realize big improvements from them.

Sad to say, not every organization is getting the best results from computing technology. Blame an organization's culture, management style, or poor choices of technology—but you can see the difference. Many businesses just don't get a payback from their investment.

So, this book has a mission (and an attitude). Its purpose is to tell it like it is. This book tells the truth about computing in business today—and how to make it better. We've left no stone unturned. We tell you about what's going on, how business changes are driving technology adoption, and what technology tools you'll need to stay ahead.

What's in This Book?

Chapter 1 introduces the major trends driving change in corporate civilization. We talk about what changes are occurring, and why change is moving at such a rapid pace. We also talk about common excuses that companies use to defend crummy computing practices. By no means do we condone them, but we accept things as they are. Your mission, as a proactive reader of this book, is to identify your own excuses and change things for the better.

Chapter 2 takes you on a whirlwind tour of technological changes. Computers—especially powerful desktop units—haven't been with us forever, after all. They remain relatively new additions to our work style. The technology (r)evolution shifted into high gear a few years ago with the invention of the microprocessor chip, and we see nothing on the horizon to slow it down.

Along with business and technology, people are changing too. Chapter 3 talks about how people are changing. Overall, the population of workers has become more computer literate. Many white-collar workers are almost as comfortable piloting a computer as they are driving a car. Moreover, the next generation of workers won't find computer literacy to be an option. Not everyone has the computer skills necessary, however, to become effective. So, in this chapter, we outline characteristics of today's corporate computer user and prescribe an action plan for making things better.

Chapter 4 is about convergence. It depicts what can happen when business, people, and technology get together. It also presents five simple steps for optimizing computing in a corporation. While we can't guarantee that these steps will yield eternal peace or perpetual corporate profits, we do know that they'll make you a lot better off and smarter.

Chapter 5 introduces two important concepts about how work and careers are changing. First, it presents the concept that companies are moving away from traditional function-led, hierarchical groups to more team- or process-oriented groups. Second, it suggests that careers are changing, too. They are no longer straightforward and predictable.

In Chapter 6, we take a look at the changing nature of business relationships. We introduce several business trends that companies are following in order to survive and thrive. We also argue the need for the computer technology that enables these changes.

Chapter 7 is the chapter of "change." In it, we talk about the impact that changes in computing make to the status quo, often changing corporate cultures as much as they change business processes. Some people are woefully unprepared for what happens—even those people who consider themselves computer professionals. We borrow two psychological models for how people deal with change and apply them to changes brought by the adoption of information technology.

Not only does change affect individuals, it affects organizations too. Chapter 8 describes the challenges to the organization. We identify a pernicious malady, which is sometimes called "mainframe myopia." The condition is characterized by information services people who wish to grasp and hold all computing power. They don't want to relinquish control, they don't want to treat the business people as internal customers, and they certainly don't want to change. We focus on how to spot the myopia and how to correct the problem before it chokes effective computing practices in the company.

Starting in Chapter 9, we turn our attention to business computing techniques, philosophy, and practices. Chapter 9 itself is dedicated to the redesign of business practices called reengineering. It defines the term *reengineer* and tells why and how information systems affect the change and lead to re-invention of business practices.

Chapter 10 discusses the technique of *outsourcing*. Outsourcing some or all of a company's computing services can be a blessing or a curse, and sometimes both.

Chapter 11 describes the reverse of outsourcing, which is insourcing: keeping or bringing work in house and controlling all resources. While insourcing is not the trend today, there are some instances where it is preferred.

Chapter 13 looks specifically at the popular technology of local area networks (LANs) and wide area networks (WANs). We firmly believe that this technology is fundamental to an effective organization in the 1990s. We will also discuss the features that make LANs and WANs valuable business tools.

Chapter 14 examines the computer-to-computer information exchanges called electronic data interchange (EDI). We review some of the issues that have prevented its widespread use until now. We describe some of the benefits companies are able to reap through the use of EDI and introduce some of the factors that may affect how it is currently used in your industry.

Chapter 15 tackles yet another important computing technology: mobile computing. We examine the basic types of mobile computing as they apply to different needs of working people.

Chapter 16 introduces technologies associated with multimedia. It discusses both the events and the catalyst technologies that are ushering in revolutionary changes in our concepts of learning and computing. We also discuss how you can take advantage of these technologies to provide a competitive advantage for your organization.

Chapter 17 defines and describes groupware—software aimed at making groups more effective. It introduces concepts of how computers can further group collaboration to boost worker and group productivity.

We turn our attention back to techniques in Chapter 18. This chapter explores new technologies and techniques that enable organizations to produce their own computing applications. We look at methods to get the business people more involved in development, as well as ways to get the computer to do most of the design and code generation. We also touch on the new paradigm in application development in which everything is an object, rather than a discrete process.

Chapter 19 explores the options available in client/server computing. This computing paradigm lets two or more computers divide processing tasks to take advantage of all the computing power available. We look at how client/server computing works and how it differs from other types of computing. We discuss the advantages as well as the caveats.

Chapter 20 closes the book with tips on how to select and hire outside help. We recognize that most organizations will not have all the expertise they need to move forward into the brave new world of computing. In this chapter, we focus on issues and techniques for securing outside help.

We've also included a couple of appendixes, and a glossary of terms, for your further research. These should serve as a handy reference.

This book is filled with golden nuggets—concepts, ideas, and tips to make you personally as well as your organization more effective with computing technology.

We hope you will enjoy our friendly approach as we tackle the issues.

About the Authors

Cheryl Currid is the founder and president of Currid & Company, a research and consulting firm that helps clients assess and apply new information technology in business. She is considered an expert in the computer industry and a visionary when it comes to picking the winners and losers of new computing technologies.

Ms. Currid's systems experience covers all facets of applying information technology to business. She has successfully implemented systems in the areas of financial, sales & marketing, manufacturing, office automation, and decision support. Each project included reengineering of business processes and deployment on a client/server or LAN platforms.

Prior to founding Currid & Company, Ms. Currid was in charge of applying information technology at a division of The Coca-Cola Company. She engineered planning, design, and implementation of an enterprise-wide network that connected virtually all computers in the company. This network served to replace mainframe processing for most business applications.

Ms. Currid is widely known throughout the computer industry for her straight talk—speeches, writing and commentary. She has been a keynote speaker at major industry events and is an international lecturer on both technology and management topics. She is an active advisory board member for industry conferences and trade shows such as articles, and writes regular columns for *InfoWorld*, *WINDOWS Magazine*, and *Network Computing*. Her opinions are sought after for analysis of computer industry trends by numerous business publications, including: *The Wall Street Journal*, *The New York Times*, *Business Week*, *Reuters*, *Associated Press*, *Investors Business Daily*, *Forbes*, *Fortune*, and *ABC News*.

Linda Musthaler is Vice President of Research with Currid & Company. She focuses on specific technology adoption issues. Her work has been published in *Computerworld*, *Data Based Advisor*, *LAN Times*, *Network World*, and several other periodicals. Ms. Musthaler is also a frequent speaker on integrating technology into the workplace. She has been an information systems professional with a successful career in large corporate environments and the public sector.

Frank Ricotta is currently a Senior Consultant with The DMW Group, Inc., a Telecommunications Architecture and Network Integration firm headquartered in San Francisco, California. The DMW Group specializes in helping organizations implement high impact technologies in order to achieve a competitive advantage within their industry. Mr. Ricotta has over 10 years of experience designing and implementing leading edge communication networks and computing environments.

Part I

The Winds of Change

Chapter 1

Changes in Business: The Major Trends

Effective use of computers in business today is an evolving art, not a science. The formula that worked for you or your company yesterday might not be so effective today—and could be dead wrong by tomorrow.

If you take nothing else away from this book, please take that thought. Write it down ten times, tattoo it to your forehead, and tell everyone you know. We hope it will either scare you or motivate you into taking action.

As you venture down the pages of this book, we hope you'll take away lots of ideas for action. No matter who you are, a business person, a corporate executive, an information services professional, or the company janitor, you should make information your business. You should know the options as well as the consequences. Sure, business computing today is confusing, complicated, and chaotic. But, it is also compulsory. Few organizations anywhere on planet Earth can operate without computers or communications. As we move forward into the next century, and business becomes more knowledge based, we firmly believe that computing strategies will become a competitive issue. In some industries, they already have. Those who can move faster and in the right directions win—but it takes information to know which way to move.

This book isn't a regular computer book—it isn't going to teach you how to calculate a sum on a column of numbers in your spreadsheet, nor will it tell you which key to push to make the spell-checker work in your word processor.

Instead, this is a book on how to make computing technology work for business—your business. We'll talk about how to rethink what computers should be doing for you and how they can be used effectively. In these pages, we'll draw on the expertise of both business and computing gurus and fuse together a philosophy about how to make information technology work for business.

We'll start by acknowledging that today we don't live in a perfect world. Many of us don't come from companies where information technology is fulfilling its promise. Then, after we make excuses and figure out whom to blame for the mess that most companies live with, we'll start talking about how and what to fix.

Change as the Only Constant

The single-syllable, six-letter word *change* is the central thought for the first part of this book. We talk about change in business, technology, and people. In this chapter, we focus on changing business dynamics. No matter which way you turn, in the business world today, change is the only constant.

We'll continue working with the *change* theme in our early chapters as we explore what's happening in technology and how people—computer users themselves—are changing. Sometimes we think all the dynamics together act like a three-ring circus—with business, technology, and people occupying a ring apiece, each performing its own show. Looking at it from the grandstand, sometimes the show looks coordinated, and at other times it doesn't.

Once we move through the issues of change, we'll start weaving our solutions. Remember, this is art and not science. We'll begin to paint the strategies for effective computing for companies. We will depart from the classic mentality that holds that computing is an extension of bean counting and begin to talk about strategies that provide information weaponry. In Part II of this book, we'll focus on how strategies like reengineering, outsourcing, insourcing, and downsizing/rightsizing computer applications affect both people and organizations. We'll even be so bold to talk about corporate politics and how that can help or hurt computing strategies. You may (or may not) be surprised to learn how much a company's political environment affects the success of its technology.

Then, we'll move to tools and techniques for delivering information effectively. We'll talk about the philosophies and technologies that work and some strategies on how to embrace them. Paychecks aside, there is a lot more to effective business computing than a payroll system. Often companies get bogged down in managing the day-to-day activities of back-office computer applications and never see the light of advanced applications.

The sad truth is that many, many companies have mediocre methods of using computers. For one reason or another, they've succumbed to suboptimization.

Excuses for Crummy Computing

Most business people have a long list of excuses for such crummy computing practices. At times, some of the excuses seem understandable, even though they rob organizations of benefits. Many organizations grow or change so fast that they neglect to keep their information systems current. Management is too busy keeping an eye on the business and lacks the time, energy, or expertise to expend the effort on computing technology.

High on the list of excuses are "we don't have time," "we can't afford to be state of the art," and "we're not sure we really have to—after all nobody has ever proved information technology promotes competitive advantage."

In some business cultures, crummy computing practices have become a time-honored, albeit time-wasting, habit. They are ingrained into the very business processes. Recently, when working with a major company, we learned management had a 17-step procedure for paying an invoice. Nobody ever questioned the process, much less the computer system that supported it. It was accepted—just the way things were, and the computer system merely reinforced a process that was cumbersome.

For the sake of argument, let's take a few of the excuses and talk about them. Some discussion will shed a light on them.

We Don't Have Time

Many people blame *time* for crummy computing. They say they're so busy doing the everyday job of business, they don't have time to consider a change. While we concede this is a hard objection to overcome, the failure to change may waste more time than it saves.

Nearly every company has time-wasting processes or procedures. Frequently these are associated with routine business processes and become ingrained into the culture of an organization. Procedures can take the form of the 17-step accounting process we just mentioned, or the 11-person sign-off procedure before a purchase order can be granted, or some other common task.

The time excuse can be tackled in small steps. Rather than undertake a huge project of business process reengineering, divide and conquer. Rather than attempt a cataclysmic change, try moving one step at a time. You can chip away at the problems.

We know of one company that simply asked its workers to submit their personal lists of time-robbing procedures. Then, a small team selected a few of the easy-to-fix practices and began identifying how new computer technology could help. They piloted a few solutions and quickly saw the benefits. Before long, the company established an improvement process, acquired a flexible computer platform, and began to tackle many problems. They started small but began to see big benefits.

We Can't Afford to Be State-of-the-Art

After time, money is the most popular excuse for inaction. People believe that replacing a computing platform will cost them money, and lots of it. Sure, there is some basis in fact for this fear, but often the new solution is so cost effective that it pays for itself quickly.

Often, the money excuse is completely without merit. New computing technology has almost followed the price curve of the pocket calculator. The price of computing power has plummeted in recent years. For example, according to a Harvard Business Review article in 1991, the cost of delivering a million instructions per second (MIP), a common unit of computing power, has declined by many orders of magnitude in the last fifteen years. The cost of delivering this power was $250,000 in 1980, $25,000 in 1985, and $2,500 in 1990. Then, costs capsized again in 1992 and 1993, when personal computer manufacturers began a price war, slashing the prices of PCs by 50–70 percent. By 1993, the cost of a MIP shot below $100 for the first time.

Even if you don't know what a MIP is or does, you can appreciate that, whatever the product or service, a price fall from $250,000 to $100 in 13 years has to have a profound effect. In this case, the effect is to make computing power very cheap and very available for people with business applications to use. We believe the real challenge isn't money, it is how to use the technology effectively.

We're Not Convinced

This excuse is popular among legions of business people. There is a vocal group of nonbelievers in technology who feel that information technology doesn't help a company gain or sustain any advantage.

Nonbelievers point to companies that have spent millions (or sometimes billions) of dollars only to be beat out in the market by more nimble, less technology-rich companies. This is called the *productivity paradox.* That is, companies spend plenty of money on technology but don't necessarily become more productive.

There are two answers to the productivity paradox. First is the measurement of productivity. Unfortunately, productivity is becoming more difficult to measure the old-fashioned way. As companies move from a manufacturing mind-set into a knowledge-based world, people can't use old methods of measurement.

While it was easy to understand the effects of applying technology when we could measure productivity in terms of "things produced," such as the number of widgets made per hour, knowledge-based work has to be measured differently. Few companies ever count the decisions per hour that their managers make. Besides, it would be a meaningless measurement. Yet, there is general agreement that a manager armed with the right knowledge about his or her business can make better decisions.

The second answer to the productivity paradox acknowledges that, yes, companies have wasted money on information technology. That is because they made one of two fatal mistakes. Either they bought the wrong technology for the job, or they bought the right technology and installed it but did not implement it. Far too often, people in business environments don't take the time to learn what computer technology can do, or they don't change their business habits to capitalize on technology. So, they buy the wrong tools for the job, or they never learn about how to use the tool properly.

Whom Are You Going to Blame?

What we see is not a pretty picture. The excuses we mention are real and repeated from company to company.

There is a crisis these days, both in large companies and small ones, about how to effectively use computing technology on which we've spent billions and billions of dollars. Few people have figured out the magic formula for getting big returns on computing investments. All too often, the benefits gained from putting computers inside of companies have failed to compute. They just don't add up.

Casting the Culprits

We can talk about all the reasons why. We can point our fingers and shake our heads in any number of directions. A popular place to put blame is squarely on the heads of corporate computer people. These are the folks that live in corporate information services (IS) departments—who devote careers to the care and feeding of big-iron computers housed in specially powered, climate controlled rooms, affectionately called glass houses. IS staffers frequently come in dead last on any popularity poll of internal corporate service providers. Why? Because nobody is happy with computing services.

Another easy target for blame is people in financial services departments. Frequently, the organizational bean counters clog the computer procurement mechanism. They use a cost-accounting mentality and require justifications that make sense by the light of the financial sun, whether or not they make technical sense. For example, we know companies that spend a lot more management talent and time to justify a color computer monitor than the monitor itself costs. We call that penny-wise, pound-foolish purchasing.

We can also point fingers at business people themselves. People in business have either shrugged off personal responsibility to learn about computing technology or jolted to the other end of the scale and become technology zealots. The zealots tend to overuse computers and develop such complicated uses that nobody can follow in their footsteps.

Companies today are still filled with people like "Teddy Too-Cool-For-School," white-collar professionals who ascended up the career ladder fast enough that they didn't have to become computer literate along the way. Now, Teddy has subordinates who can push the keys for him. Teddy may be a great guy, but he doesn't have a clue about what computers can really do. He is also somewhat a hostage to his subordinates, and he sometimes gets filtered information. Since Teddy can't find things for himself, he knows only what other people want him to know. Teddy can be dangerous when it comes to making decisions about corporate computing. He also has a problem properly calibrating his expectations of computing technology; he expects either too much or too little of technology.

In other cases, we see the super-duper knowledge brokers like "Steve Splash" who can serve up a solution in a second. Steve and his buddies are almost as dangerous as Teddy—but in an entirely different way.

These people have become genuine computing rocket scientists. They can make their personal computers do just about anything—formulate fantastic financial forecasts, generate great graphs, deliver dazzling data, you name it. Their single, but fatal, flaw is that they expect everyone else (including all mere mortals) to be in the same league. Unfortunately, even though these splashy super knowledge engineers can make lots of great computing feats happen, they also tend to build quick and dirty solutions that nobody else can maintain. Often, the benefit of their work is short lived, because if (and when) they move into another job or leave the company, no one can figure out how to keep things going.

Finally, we can blame "Edwin the Executive." Edwin, or Mr. E., hasn't articulated a clear mission for the company. He, and the company direction, are constant moving targets. His opinions and his management practices, philosophies, and structures change like the weather. While this is common in companies today, it plays havoc with information systems.

Understanding All the Characters

We'll talk more about the effects of constantly changing management priorities throughout the first part of this book. Realistically, we aren't ready to pin all the blame on Edwin, Steve, Teddy, normal business people, or the IS staff. We think each constituency is guilty as charged, and we'll talk later about strategies and self-help ideas for each group.

Understanding the Art of Building a Computing Strategy

Rather than bemoan the fact that change keeps upsetting our business computing boat, we'll instead acknowledge it and figure out how to develop effective computing strategies to ride the waves of change.

And, as you may have guessed, our ideas involve more art than science. We won't give you a precise formula to follow, but we will give you the pens, pencils, brushes, and crayons to create your own picture.

We sketch out our thoughts for the future by citing a number of the management gurus like Peter Drucker, Tom Peters, Peter Senge, Stan Davis, and a few others. While each of the modern-day soothsayers offers a unique perspective on the past, present, and future of business, we think they all have some merit.

Here are a few trends that we have gleaned from the management gurus:

- Organizations are becoming knowledge based.
- Knowledge in many disciplines has a shelf life.
- Globalization will drive technology adoption.
- Workgroups are changing.
- Virtual corporations are forming.

Organizations Are Becoming Knowledge Based

Nearly every management guru is purveying the notion that business today is moving from a "thing"-based to a "knowledge"-based economy. Peter Drucker's September 1992 *Harvard Business Review* article, "The New Society of Organizations," points out the fundamental shift between the old-model business structures and the new one. He says:

> In this society, knowledge is the primary resource for individuals and for the economy overall. Land, labor, and capital—the economist's traditional factors of productions—do not disappear, but they become secondary. They can be obtained, and obtained easily, provided there is specialized knowledge.

Clearly, Drucker places the importance of knowledge above things we can measure. You can easily buy bricks, buildings, or other hard assets—but you need knowledge to buy and use them wisely.

Other industries, such as manufacturing and retailing, are increasingly becoming knowledge based. People in these businesses need to know how much to make or how much to buy. While the product itself is important, information about the product is equally important. It doesn't do any good to make the best purple tennis shoes on Earth if nobody wants to buy purple tennis shoes.

Knowledge in Many Disciplines Has a Shelf Life

Another important concept is shelf life. Knowledge can quickly become outdated in this new business world. Consumer trends change, competitors enter and leave markets, and paradigms shift. What we knew about business yesterday might not be so meaningful today.

More and more companies are finding that knowledge must be renewed. It is important for them to spot historical trends and then to project whether these trends are valid for the future.

Globalization Will Drive Technology Adoption

More and more companies find themselves in global markets. This is both good news and bad news.

The good news of globalization is that companies find new markets for their products and services. No longer does a firm find that its geographic borders have to become economic boundaries. As a result, organizations both large and small quickly find themselves selling more outside of their homeland than in it. The Coca-Cola Company, for example, is said to derive more than 80 percent of its operating revenues from outside the United States. Compaq Computer achieves over 60 percent of its sales from beyond the continental 48 states, and even small companies find the same scenario.

There is bad news to globalization too. Not only do you find new markets for your goods and services, you find new competitors too. Today, a competitor isn't just the company next door, it can be based anywhere on the globe.

It is said that many companies based in Japan, which have long amortized the benefits of their superior manufacturing abilities, are now looking for new and better business practices to compete. They now look to information technology to help them become more competitive.

In fact, information technology may well be the next way that companies can compete. For those who can churn raw data into information and then act on it quickly, a new competitiveness emerges.

Workgroups Are Changing

There are two interesting trends in workgroups that have emerged in the 1990s. Workgroups are no longer geographical, nor are they function-based.

More and more companies are finding that, when they use the right computer and communications gear, they can form workgroups based on talent and not geographic convenience. Accounting firms, engineering firms, and other knowledge-based organizations are now selecting members for a workgroup on the basis of talent and expertise. Sometimes that means pulling together teams that span from Tokyo to Toledo.

A second new phenomenon is how companies are beginning to adopt cross-functional teams. Instead of following processes through normal functions like accounting, manufacturing, and marketing, managers find they can draft representatives from different functional groups and place them into mission-oriented teams. We've heard them called Noah's Arc teams because they draft two from each species—that is, two from sales, two from finance, two from manufacturing, and so on. Then, these people are formed into teams, given a goal, and set off to discover the new world—or to develop a new product, as the case may be. We'll talk more about the effects of workgroup changes in Chapter 5.

Virtual Corporations Are Forming

Still another trend emerging in business today is the formation of virtual corporations or virtual teams. In this case, organizations bring in talent simply to deal with the task at hand, and not necessarily to stay forever.

In virtual corporations, people from the outside come in to lend specific expertise that the company needs now. When the project is over, the experts go away.

These virtual groups are becoming very popular but, once again, require effective information technology to work well. It takes very flexible systems and access to systems to accommodate new non-employee people.

Changing Business Climates

The last decade has ushered in some profound changes to the business climate that present many challenges to organizations large and small. As we have mentioned, the world has gotten smaller, and the globalization of markets has forced companies to become more competitive or face possible extinction. The privatization of government-owned businesses has proved challenging to countries in transition like Great Britain, Mexico, Canada, and the former Soviet Union, but it is also making them more globally competitive. Corporations feel the pressure to get new products and services to market in a shorter time. Moreover, the recent spate of mergers and takeovers has caused tremendous upheaval in the business environment for many companies. These and other changes present challenges to managers at all levels.

Many successful companies are using information technology (IT) to respond to the challenges. John Rockart and James Short of the MIT Sloan School of Management suggest that business needs drive increased information technology capability, which in turn creates additional business needs. Robust information systems allow a company to adapt quickly to changing business needs. These systems emphasize placing "information" (as opposed to merely "data") into the hands of the line of business managers, allowing decisions to be made closer to the customer. More and more frequently, companies are turning to workgroup computing and "downsized" computer systems to respond effectively and inexpensively to dynamic business demands and to survive competitive pressures. These systems are frequently far better suited than their predecessors to handle dynamic business changes because they localize processing power. They allow people to apply computing power specifically to selected tasks.

We believe there are several basic business trends that have changed the way companies operate. They include:

- Merger Mania
- Employee Empowerment
- Globalization
- Time to Market
- Privatization
- Competitive Pressures

Merger Mania

In the 1980s, corporate America went on a binge of takeovers and mergers. According to Alvin Toffler's book, *Powershift,* "...of the 20 largest deals in business history, all consummated between 1985 and 1989." In 1985, General Motors bought control of Hughes Aircraft, paying (at the time) the largest amount ever for a corporation—$4.7 billion. That buyout was followed by a flock of other mergers. In 1988, there were nearly 3500 acquisitions or mergers—amounting to an astronomical $227 billion. Then, in 1989, all old records were shattered again, as RJR-Nabisco was taken over for $25 billion.

We've also seen other unions as companies go global and cross national frontiers, such as Japan's Bridgestone acquiring Firestone, England's Cadbury Schweppes taking over France's Chocolat Poulain, and Sony buying Columbia pictures.

There are other types of relationships—in the shape of strategic alliances, joint ventures, or partnerships. Companies are reaching out to co-market or co-manufacture goods or services, trying to share the talent and effort, the risks and rewards of getting new ideas off the ground.

Table 1 lists some examples of trends of corporate mergers and takeovers. The list includes buyouts and companies that have resisted takeovers. It will make you aware of the intensity of corporate warfare.

TABLE 1 Representative Corporate Mergers and Acquisitions

INDUSTRY	EXAMPLE OF TRENDS TOWARD MERGERS AND TAKEOVERS
BANKING	Bankers increasingly view consolidation as a way of paring costs and eliminating wasteful overcapacity.
	The merger of Chemical Banking Corp. and the Manufacturers Hanover Corp. in New York is an early round of a banking merger mania.
	C&S Bank and NCNB merge to form Nations Bank.
	First Bank System, the second-largest banking company in Minnesota, announces November 9, 1992 that it has reached an agreement to buy Colorado National Bankshares for about $500 million in stock. The fate of Aachener und Munchener, Germany's second-largest insurer, and BfG, its banking subsidiary, is up in the air. Two French companies, Assurances Générales de France and Credit Lyonnais, are interested in buying them, but stockholders are upset at the offering prices.
ADVERTISING	Kenneth Roman, former chairman and CEO of ad agency Ogilvy & Mather, wins more than $1.5 million from WPP Group in an arbitration decision that ends a long dispute arising from the hostile takeover of Ogilvy.
LUXURY FASHIONS	*The Independent,* a British newspaper, reports November 1, 1992 that Dunhill Holdings PLC, the British maker of luxury goods, is negotiating a takeover of Guccio Gucci SpA, the Italian fashion house.
NEWSPAPERS	When a bankruptcy judge voids the lifetime job guarantees for 167 employees of the *New York Daily News,* a major obstacle is removed in Mortimer Zuckerman's takeover of the paper. Zuckerman still faces unresolved conflicts with two of the paper's 10 unions. *Izvestia,* a Russian publication, is formally acquired by its staff but continues to use government-owned premises and equipment and to rely on government subsidies. It is also the target of a government takeover.

TABLE 1 Representative Corporate Mergers and Acquisitions (Continued)

INDUSTRY	EXAMPLE OF TRENDS TOWARD MERGERS AND TAKEOVERS
BAKING	Tomkins PLC, an Anglo-US conglomerate, offers £952 million ($1.5 billion) in cash or a mixture of cash and shares for UK breadmaker Ranks Hovis McDougall PLC, topping rival Hanson PLC's hostile £780 million offer. RHM accepts the Tomkins offer.
FOOD PROCESSING	Anglo-American conglomerate Hanson PLC launches a hostile takeover of Ranks Hovis McDougall (RHM) on October 5, 1992, with a £780 million ($1.35 billion) cash bid. Analysts expect Hanson to sell most of the company's major brands and concentrate on RHM's baking and flour-milling business. RHM's board promptly rejects the cash offer.
AIRLINES	Two Australian newspapers reports October 28, 1992 that British Airways PLC is considering a takeover bid for Australia's Qantas Airways that would involve Australian private investors. The takeover is reported to be worth at least $1.8 billion.
HYDRAULIC PRODUCTS	Although Applied Power Corporation's 1989 acquisition of Barry Wright Corp. has not yet paid off, the maker of hydraulic products may be in the early stages of recovering from a slump that was aggravated by the costly takeover.
PENS AND PENCILS	Gillette Co.'s proposed $562 million purchase of Parker Pen Holdings Ltd. of Britain comes under increased scrutiny by British regulators over antitrust issues. British Trade and Industry Secretary Michael Heseltine asks the British Monopolies and Mergers Commission to investigate and recommend whether the takeover should be allowed.

Employee Empowerment

More and more senior managers are finding that the best decisions are made by people on the front lines of an organization, not those inside a corporate tower. It has become a popular management notion to empower employees to make decisions. The only problem is, you can't empower employees without giving them information so they can make better decisions. In other words, you can't delegate decision making to dummies.

Also, before we can talk about employee empowerment, we have to look at another concept called *demassification*, which means decentralizing and downsizing, reducing the size of America's large corporate operating units.

Demassification is a term coined by M. M. Stuckey in his books *Demassification: A Cost Comparison of Micro vs. Mini* and *Demass: Transforming the Dinosaur Corporation*; it is also used by Alvin Toffler in his book *Powershift*.

In behemoth organizations (massification), control is all-important; consequently, managers refuse to relinquish control to workgroups or to encourage their success. But even in small organizations, there are corporate executives and managers who would not think of decentralizing control and letting employees become responsible for the quality and quantity of their production and services.

Top management tries to be the quality integrator for the company and to prescribe how to save a few dollars here or save time there. They encourage workers to submit suggestions for time and cost savings but fail to carry through the ideas by giving the workers the tools and responsibility to implement them.

Why is American industry demassifying? It has to decentralize to compete in today's global economy. Downsizing computer systems (or distributing computer power) is one of the key ingredients for successful demassification because large mainframe computing systems tend to freeze company operations, locking up data.

At the same time, corporate managers must encourage employee innovation and creativity because no product, from computers to medical plans, is safe from highly innovative competitors. Free workers tend to be much more creative and productive than those who are closely supervised. Managers who use fear tactics destroy creativity and kill potentially productive ideas. Furthermore, managers need to tolerate intelligent error because, although many new ideas prove to be unworkable, they must be brainstormed until a good one emerges. Companies must allow for experimental errors to obtain success.

Management experts generally believe that Goliath corporations' obsession with control over their employees prevents employee empowerment. It prevents any identification or sense of ownership workers have of their projects. Just when employee empowerment was most needed for corporations to become more competitive, they squashed the idea to defend themselves against takeovers.

Control from the top down doesn't work. Empowered employees produce more and higher-quality work in less time.

A case in point: At Kodak, Jack Philbin converted the top-down-control corporation culture to one where the employees maintained their department data on personal computers instead of a mainframe. They worked together in small business units. The result was raised employee morale, better communications, and the largest cost savings Kodak had ever achieved. Philbin said that demassification led to flow manufacturing and away from the control culture. Employee empowerment enabled total quality management through trust. He put the power where it was most effective—with his employees. The company's success depended on employees' voluntarily contributing to cost reductions and increased productivity.

Employee empowerment results from a different kind of downsizing, called rightsizing, which applies to a company's human resources. Because managers are fearful of failure, they are not eager to pass decision-making authority to the workers. Furthermore, the workers must grow more aware of what's happening in other departments. They need access to data to do their jobs, and they must be trained to understand the interrelationships of financial processes and operations. For example, the shop foreman may wish to order enough parts to continue production for a full year, but he needs to know that there is a high cost associated with keeping this inventory in the warehouse. The employees no longer have just one job to do; they must work cooperatively with others and understand each other's jobs as well. Management must train or hire the people who have the ability to work together, to look constantly for ways to break bottlenecks, and to maintain a sense of responsibility for handling their new power.

Our major institutional forms come from a revolutionary era that viewed humankind as equal and rational. We know that people are basically emotional in their motivations and only sporadically able to sustain the tensions and stress involved in taking thought to direct their actions. Technology and science are continually opening up new dimensions of complexity even in the commonplace situations of daily life, physical health, mental hygiene, economics, business, and government. We, however, rely upon our human rational power and our ability to recognize useful information to acquire knowledge. Furthermore, with employee empowerment, we continue to leave everything up to the individual's precarious ability to "use his head." The aim of the new knowledge era is to avoid dull, methodical culture. Instead, we want to create a cultural situation that minimizes the occasions for wasteful mistakes and that frees energy and resources for the creativeness of collective human minds.

Our culture is part of a larger world where external chaos renders internal order more difficult to achieve. These difficulties, however, should not be used as an excuse for delay in setting our corporate houses in order.

Time to Market

Massiveness often lengthens time to market. This fact flies in the face of new consumer and business desires, which are primarily: the need for speed. People will flock to anything they can get in an instant. Look at the popularity of instant photos, one-hour eyeglasses, and fast bank loan approvals. More and more, people find that time is a competitive issue. The faster the service, the better.

Computer-assisted procedures are speeding the process of making everything from eyeglasses to automobiles. For example, making automobiles a *just-in-time* process has quickened the pace of manufacturing while lowering the cost of inventories. As each part is used, the identifying card or tag attached to the part is sent forward to order a new part for the next production run. It arrives *just in time* to be used in the next time cycle. In automated just-in-time systems, the tag or card is replaced by a transaction or trigger in the computer system that issues an electronic order to the vendor through EDI (electronic data interchange). The vendors and manufacturer's suppliers use the same formats for their electronic data so that their computers and data bases will be compatible. The manufacturer's electronic purchase orders are converted to the vendor's bills, and payment can be made electronically to the vendor's bank.

Today's supermarkets, selling over 20,000 different items, use Universal Product Code scanners at the checkout counters to keep track of the sales. For each item, they can determine profitability, timing of advertising, special promotions, costs, prices, coupons, and location on the shelves. Furthermore, they can tell a lot about their customers, such as the types of food they eat and even what magazines they read.

This new knowledge accelerates response to customers' demands and lifestyles. It is driving commerce and industry to a real-time and instantaneous economy. Moreover, this knowledge power has become a substitute for time expenditure.

Table 2 lists some trends toward reducing time to market.

TABLE 2 Trends Toward Reducing Time to Market

TYPES OF SYSTEMS	EXAMPLE OF TRENDS TOWARD REDUCING TIME TO MARKET
DATA PROCESSING TOOLS	Application frameworks, class libraries, and GUI toolkits address the same basic software-development issues, including reducing coding effort and speeding time to market.
COMPUTER-ASSISTED DRAWING	The benefits of automated time compression using Computer-Assisted Drawing (CAD) and its software can add velocity to a company's manufacturing capability and reduce time to market.
TESTING SOFTWARE	GenRad Inc. plans to revive a "classic" test system with the 179X Platinum for running programs written for the 1970s era. GenRad's primary marketing thrust for the 1990s is the reduction of customers' time to market.
SPECIALIZED SOFTWARE	New software allows engineers to provide quick electronic product thermal analysis, increases reliability, and cuts time to market. Efficient product cooling patterns can be designed by engineers right from the beginning.
FILM	The success of overseas films in Montreal has weakened by as much as 60 percent in the past five years. Quebec used to be the only territory in Canada where U.S. films didn't dominate 97–98 percent of the screen time. The market share for overseas fare slipped when U.S. majors began releasing French-dubbed versions of U.S. pictures closer to the date of the original English versions.
PRODUCT DESIGN AND DEVELOPMENT	Engineers and managers should resist quick time-to-market schedules. A prudent manager will make time for proper design and quality. It is important to design properly the first time. Companies that are able to turn out products more quickly than ever before might be well served by giving some of that time back to the designers.
	Orion Research Inc.'s experience with concurrent engineering partnerships between supplier and manufacturer have benefited the design and production of injection-molded parts by cutting time to market. These partnerships produce better designs of higher quality.

TABLE 2 Trends Toward Reducing Time to Market (Continued)

TYPES OF SYSTEMS	EXAMPLE OF TRENDS TOWARD REDUCING TIME TO MARKET
ELECTRONIC DATA INTERCHANGE	The types of applications and approaches businesses take to improve the quality of goods and services affect suppliers by placing new demands upon them and offering them new opportunities. It is important to design properly the first time.
MANUFACTURING	The Japanese agile manufacturing technique called lean production slashes the fat by starting with inventories of finished goods and following through with just-in-time (JIT) manufacturing. The agile manufacturing system has stood the traditional U.S. approach on its ear.
JUST IN TIME	Several small manufacturing firms were surveyed to examine the benefits and problems associated with implementation of the just-in-time (JIT) philosophy at customer and supplier linkages. Significant reductions in inventory were reported after JIT implementation. The decision of Cybex, a Long Island, NY, manufacturer of fitness products, to implement just in time attempts to eliminate waste in time, labor, and inventory.
FLEXIBLE SPECIALIZATION	For many years, industrial mass production was criticized for its consequences on conditions of work. Today, this mode of production is facing another crisis caused by its inability to respond to changing market demands. A new mode of production called flexible specialization may be an answer to both these crises, while just-in-time production may be a solution to the required responsiveness of a firm. Just-in-time business documents are impractical without an efficient, competitive way of transporting them. Electronic data interchange can benefit U.S. companies' competitiveness by helping to speed the products through ports. The just-in-time concept of manufacturing needs to be seen as an organizational philosophy.

TABLE 2 Trends Toward Reducing Time to Market (Continued)

TYPES OF SYSTEMS	EXAMPLE OF TRENDS TOWARD REDUCING TIME TO MARKET
AUTOMATED DOCUMENTATION SYSTEM	Texas Utilities Electric Co. was faced with burgeoning files of documentation required for the construction and operation of a nuclear facility, so the company replaced its personnel-intensive document storage and retrieval system with a computer-based system. The system is establishing itself as more than just a time-saver.
CHEMICALS PRODUCTION AUTOMATION	The Butyl Polymers Americas (BPA) unit of Exxon Chemical Co. was recently awarded one of the four 1991 Shingo prizes for excellence in manufacturing. They achieved the awards by using just-in-time (JIT) and single-minute exchange of dies (SMED) techniques.

Privatization

In Russia, Canada, Great Britain, Mexico, and several European countries, governments are selling off public enterprises. In the United States, Total Quality Management has become a government practice in an effort to remove power from specialists and managers who control information. Governments increasingly are farming out contracts to private industry through competitive procurements.

Privatization has become the dominant trend irrespective of whether a government is communist or capitalist, liberal or conservative. It's comparable to the push in private industry to flatten hierarchical structures. What has become apparent is that, when we change the structure of business and industry, a chasm appears between the private sector and the government. So, government is adopting corresponding changes: Japan has privatized its railroad; Britain has given up the management of its major airports to private industry; Germany has divested itself of Volkswagen; and France has privatized Matra, St.-Gobain, Paibas, Compagnie Générale d'Electricité, and Havas.

Privatization is not a panacea, but it does relieve a government of some of its bureaucracy. More important, it's accelerating the response to consumers of both the governments and the new private companies. Congress has been facing the issue of privatizing its public information. Today, anyone with a desktop computer can access government databases on hundreds of topics. Some agencies want to contract the distribution to private firms who would sell the information. Librarians, researchers, and non-profit companies want the information to remain public.

Authors like Alvin Toffler and M. M. Stuckey say that privatization thus far is just the tip of the iceberg. This trend will grow into the 21st century. Table 3 lists some examples of the continuing worldwide move to privatization.

TABLE 3 Trends of Privatization

GOVERNMENT OR ISSUE	EXAMPLE OF TRENDS IN PRIVATIZATION
EASTERN EUROPEAN	Economic transformations attempted in East Central Europe in 1992 have been marked by "path dependence." European countries are using privatization strategies to decide questions such as how to value state assets. Even without competitive pressures, private ownership seems to work better than public management. Privatization trends are rapidly taking place as socialist governments are being dismantled in Eastern Europe.
HUNGARY	Hungary's grappling with the monumental task of privatization involves the relationship between ownership and the entrenched authority structure. This is the main problem of political economy.
GOVERNMENT SERVICES	Although labor leaders believe cost savings from privatization come from reduced wages and worker benefits, it is likely to remain a popular option for governments that want to continue to maintain government services without raising taxes. Some government leaders debate whether privatization improves government services and lowers costs. One asserts that privatization spurs competition, innovation, and efficiency, and another argues that privatization can lead to contractor manipulation.

TABLE 3 Trends of Privatization (Continued)

GOVERNMENT OR ISSUE	EXAMPLE OF TRENDS IN PRIVATIZATION
ITALY	Starting Italy's privatization program, the state-owned Istituto per la Ricostruzione Industriale has decided to sell its 67-percent stake in Credito Italiano, the country's fifth-largest bank, through an auction.
RUSSIA	In an interview, Viktor Golubev and Georgy Tal, both experts in Russia's Supreme Soviet, discuss their participation in the development of the state privatization program. The "collective" version of privatization appears to be the version being accepted by the majority of workers.
	The city of Nizhny Novgorod, Russia, is introducing economic reforms that are doing well in its market-economy reform efforts. It has been successful because of its long business tradition and a go-ahead local government.
GERMANY	As the outlook for eastern Germany's economy becomes further depressed, privatized companies are becoming less private—and those still owned by the state are more likely to decline further. The privatization of the eastern German steel industry, which pits explosive regional politics against grim economic realities, is proving to be one of the most tortuous chapters in Germany's industrial unification. The steel industry once employed 70,000 people in the region, but sinking market demand and plunging prices provide scant economic justification for saving eastern Germany's inefficient and antiquated steel producers.
POLAND, HUNGARY, AND THE CZECH REPUBLIC	The privatization of large and medium-sized state-owned enterprises in Poland, Hungary, and The Czech Republic has been characterized by varied approaches and results.
PHILIPPINES	Philippines President Fidel Ramos orders the government to accelerate sell-offs of entities or activities already scheduled for privatization. So far, 74 of the 122 government enterprises marked for privatization have been sold for a total of 22 billion pesos ($916.7 million).

TABLE 3 Trends of Privatization (Continued)

GOVERNMENT OR ISSUE	EXAMPLE OF TRENDS IN PRIVATIZATION
INDIA	India's sales of stock in state-owned concerns to financial institutions and private investors, part of the economic reforms begun in June 1991, are gaining momentum. The sale of minority stakes is not a move toward privatization, however.
UNITED STATES	Significant changes and improvements have occurred since Operations Management International, a private contractor, began operating Hinesville, Georgia's public works department.
GREAT BRITAIN	The UK government is making plans to privatize Scotland's water; however, a poll shows that 89 percent of Scots oppose the privatization of their water.
PORTUGAL	Portugal's Radio Comercial, the state-run radio network, will be sold on Lisbon's stock exchange in a three-stage privatization.
FRANCE	The French government will sell at least six million, or about 10.5 percent, of its ordinary shares in state-owned chemical and pharmaceutical company Rhone-Poulenc SA and will perform a share swap for non-voting certificates. The sale could yield about three billion francs ($572.4 million).
ARGENTINA	Argentina's National Atomic Energy Commission says the country intends to privatize its nuclear program, which includes two operating reactors and a third to start operating in 1996, as well as plants to enrich uranium and produce heavy water.

Competitive Pressures

Throughout the United States, companies that failed to respond to early warning signals of the shift from manual labor to service work were forced into massive layoffs and bankruptcies. They were late installing computers, and they failed to incorporate electronic information systems to keep ahead of their competitors. Some placed the blame on the competition from foreign trade. Others came up with excuses about interest rates or government regulation. The truth is, often their competition beat them by adopting better operating practices and better information technology.

In order to compete successfully, new products must be brought to market quickly. Companies need an infrastructure of electronic data flow and communications, and they need to train their staff to use this information network effectively. To compete globally, businesses must recognize the need to educate their employees about their infrastructure.

Some interesting trends of worldwide competitive pressure are listed in Table 4.

TABLE 4 Trends In Competitive Pressures

INDUSTRY	EXAMPLE OF TRENDS IN COMPETITIVE PRESSURES
UNION VS. NONUNION	An examination of effects of unions on employment at the plant level shows that employment grows about four percentage points per year more slowly in union than in nonunion plants. This suggests that unionized plants face substantial competitive pressure.
SOFTWARE INDUSTRY	Competitive pressures on the software industry from home and abroad are causing companies to examine the effectiveness of their software development and evolution processes. Companies are using an in-process development model for software process management.
GROCERY	Kroger Co., the Cincinnati grocery chain, cites competitive pressures in key markets plus lingering effects of a 10-week strike in Michigan as the reasons for constrained earnings.
MAGAZINE	Suffering from sluggish sales and the fear of competitive pressures from *Harper's Bazaar* and *Vogue*, Hachette Magazines is undertaking a diversification of its flagship magazine, *Elle*, including more conservative fashion coverage. Competitive pressures and the explosive growth of cable TV have forced *TV Guide* to become more "cable-friendly." The magazine has increased its local cable TV listings and developed system-specific editions and cable-oriented spin-offs for use as marketing tools.
PETROLEUM BP	America, a unit of British Petroleum PLC, says it plans to eliminate 600 to 700 jobs at its facilities in Cleveland, where 3,500 people are employed. BP cites the recession's impact on oil prices and strong competitive pressures as reasons for the cutbacks.

TABLE 4 Trends In Competitive Pressures (Continued)

INDUSTRY	EXAMPLE OF TRENDS IN COMPETITIVE PRESSURES
FOOD, HEALTH CARE, AND CONSUMER PRODUCTS	First quarter 1992 earnings for several food companies were lower due to competitive pressures and higher marketing costs. Results were also lower for health care products, medical supplies, pharmaceuticals, cosmetics, household products, and consumer products. Wal-Mart places competitive pressure on their suppliers by demanding that the suppliers fill all orders with 100 percent accuracy as to the quantity or amount, sizes, and models of products.
SERVICE INDUSTRY	Several companies are making a regular practice of offering executive briefings and educating customers about the services they offer on an industry-by-industry basis. In most of these cases, the customers get a lot of sizzle, made necessary by competitive pressures.
	Intense competitive pressures are driving all kinds of businesses to a variety of close working relationships that require foundations of trust. Synergism and mutual benefits are received when family enterprises demonstrate their commitments to venture partners or to the communities where they do business.
INSURANCE	American International Group has far outstripped its rivals in the insurance business. It has achieved an impressive record, but it seems destined for much slower growth in the years ahead because of competitive pressures. Diversification risks seem to be catching up with the firm.
SPORT-UTILITIES, MINIVANS, AND SMALL CARS	European product planners have turned their sights toward pleasing U.S. customers. They are exploring several new niches, including sport-utilities, minivans, and small urban cars.

Business Challenges in a Nutshell

No matter which way you turn, no matter what direction you look, you'll see that the business of business is in a constant state of change. Fortunes are won and lost depending on how smart people in companies can deal with business changes.

Business, however, isn't the only thing changing. So is information technology. In the next chapter we'll examine some of the changes in technology and how the right answer continues to evolve.

Chapter 2

Changes in Information Technology

It was once quipped that:

> If the automobile industry had progressed as fast as the computer industry, we'd all be driving Rolls-Royces that go a million miles an hour and cost 25 cents.

Fasten your seat belt, grab your helmet. The changes in information technology make for quite a ride. Even if you are only a casual observer of the computer industry, you can't help but notice the breathtaking improvements in information technology, in hardware, and in software.

In this chapter we'll talk technology—but don't let that scare you. Our aim is to set the stage: to give you a grounding in where things are today, and where they are going. Computer technology has changed drastically in the brief history of computing in business. While computers have been around since the 1940s, the first real commercial business applications didn't present themselves until the early 1960s.

Computer users have changed as much as the technology. Once the domain of specially trained technicians, computers now are available at inexpensive prices and sold at discount stores. Moreover, technology will continue to change as processor and data storage improvements race ahead at a breakneck pace.

Predictions and Promises

Unlike other industries, the computer industry has benefited from some very accurate predictions.

In the 1960s, one of the industry's most visionary leaders expressed a wild expectation of how fast and far the technology would extend itself. It is called *Moore's Law* and has proved incredibly accurate for well over thirty years. Gordon Moore, one of the founders of Intel and now its Chairman, postulated that microprocessor chip capacity would double every two years. Wiser and more learned engineers of the day scoffed at the wild-eyed enthusiasm of Moore's projection. Surely he couldn't be right. In 1976, it is true, Moore had to revise his "law" to state that capacity would double every 18 months. You see, the technology started moving even faster than his most optimistic projection had suggested.

Power is accelerating, and prices are plummeting. The process of development is resulting in mass-produced computing machines, sometimes called servers, that are quickly taking the place of large mainframes. This process, called downsizing, is having a big effect on how people use computing technology in organizations. We predict those trends will continue.

Recently, Compaq Computer's Executive Vice President and technology visionary, Gary Stimac, predicted likely advances in PC-based server computers. His chart, shown in Table 5, shows the rapid changes in microprocessor-based technology.

The table clearly shows the evolving curve of power versus price.

Other computer industry experts agree with Stimac. As for processing power, NCR Senior Vice President and chief scientist Philip M. Neches claims that, in each 3½-year generation, microprocessor performance quadruples, while mainframe performance merely doubles. With every generation, microprocessors take two steps toward mainframes in terms of absolute performance. Says Neches, "A decade ago, the gap between mainframes and microcomputers was very large; today it is almost gone."

We see the revolution in computing power more as evolution—especially as it is coming to pass in business today.

TABLE 5 PC Server Hardware Directions

Performance Area	1989	1993	1997 (est.)
Database Performance	45 tps	350 tps	2,000 tps
File I/O	200 IOPS	1,500 IOPS	6,000 IOPS
Storage	2GB (2,048MB)	50–100GB	up to 1TB
List Price	$25,000	$20,000	$15,000

Tps = transactions per second, a standard measurement of the speed with which transactions can be processed by the database

IOPS = indicator of file inputs/outputs (reads, writes) per second

GB = gigabyte (1,024 megabytes, or about 1 billion characters)

TB = terabyte (1,024 gigabytes)

Source: Compaq Computer Corporation

Darwin and Corporate Computing—The Evolution of Technology

For midsize and large organizations, computing has undergone four distinct eras in the last four decades. Interestingly enough, the eras tend to start at the midpoint of the decade and last for about ten years.

We see the computing eras as having followed machine evolution in the past, but this pattern may not persist into the future. Instead, we believe, by the year 2000, computing will be centered around people.

Our evolutionary view of computing recognizes these eras:

The 1960s: The Mainframe Era
The 1970s: The Minicomputer Era
The 1980s: The Personal Computer Era
The 1990s: The Network Era
The 2000s: The People's Era

The 1960s: The Mainframe Era

The mainframe era was at its height in the 1960s, when IBM and others brought commercial computing machines to market. These machines were a boon to many growing businesses.

Although they were very expensive, they could calculate numbers and store data better than Miss Nelly's accounting department.

Technically speaking, mainframe computing was very specialized and very expensive. It was necessary to place the computers in specially climate-controlled, dust-free rooms with plumbing to water-cool some of the components. Frequently, these rooms were erected with glass walls—so passersby could marvel at the machines—and came to be known as the glass house.

The operating environments and software for early mainframe computers were designed carefully to conserve expensive resources. Specially trained operators had to learn very cryptic commands to program the computers and make them perform tasks. Most business people simply had no access to the computers other than getting predefined reports from them.

Mainframe computers also forced people to share the computing power of a company. Jobs had to be prioritized and run in special sequences to optimize the use of the computing machine. This system sometimes turned into a problem, when company workers didn't want to share or felt that their jobs should come first. It was common to have to wait as many as two weeks for a report. By the time some people got the information they needed to solve a problem, the problem had passed.

Socially, mainframe computers centralized everything. For some jobs that was good, and for others it was bad. The computer took data out of the hands of business people, but sometimes it was hard to get information back. Conceptually, these computers were supposed to be more efficient, to free people from laborious tasks, as well as to remove the need to keep independent files and reports. Realistically, people did not stop keeping their own data files. Because these computers were difficult and expensive to program, people couldn't get back information in flexible formats.

On balance, the mainframe era provided more benefits to business than it took away. Clearly, few people would want to return to a world of batch processing and shared expensive resources. But, in its day, mainframe computing allowed companies to make certain tasks more automated and more efficient.

The 1970s: The Minicomputer Era

The minicomputer era was a variation on the theme of mainframes. Minicomputers, centralized processing computers smaller than mainframes, were deployed as solutions for certain departments. Usually, minicomputers were very specialized computers with specific purposes, such as centralized word processing.

For example, rather than have everyone in the company share the processing power of a central mainframe, companies would often buy minicomputers for certain applications in various departments. Perhaps the finance department would have its own minicomputer, as would the marketing department.

Minicomputers still required users to share computer processing power, but there were fewer users per machine. As long as a computer was specifically tuned for the application and specific number of users, its performance was easily acceptable. Changes, in applications or numbers of users, would cause havoc with performance.

Technically speaking, minicomputers were a little easier to operate than their mainframe counterparts. Advances in operating systems were beginning to take hold, and more user-friendly administrative tools and programming languages came to market. These advances opened computing to many new people, although specialized training was still required.

Socially, the minicomputer era was a first pass at getting information to business people quicker. The minicomputer operating environment was more oriented to online questions and answers than the mainframe environment, so people didn't have to wait for their information request jobs to be processed in long queues. This change gave end users a taste of readily available information.

The 1980s: The Personal Computer Era

The most significant architectural change in computing came with the acceptance of personal computers (PCs) in business environments. Although the first PCs were developed in the mid 1970s, the computers didn't show up in many office environments until the early to mid 1980s. Even then, seasoned information services professionals doubted that these small desktop computers had much value.

The fundamental difference between the personal computer era and the eras that preceded it was simply where the processors were located. Instead of having to share the power of a computer somewhere in a distant room, people could have computing capability right on their desks. Moreover, they didn't have to ask anyone's permission to calculate or compute something.

Another difference between the personal computer era and preceding eras lay in the concept of a multiuse computer. A PC would be a spreadsheet calculator in the morning, a word processor at noon, a graphics workstation in the afternoon, and a database engine that evening. People didn't have to use different computers for different tasks.

Socially, PCs were quickly accepted by business people as great means of empowerment. PCs were much easier to use even than minicomputers, and a multitude of software packages could be loaded on the desktop devices. This ease of use gave business people enormous power and a feeling of control over their work destiny.

Unfortunately, the PC era did have some drawbacks. Most PCs were installed as standalone devices. This fact made sharing information difficult. If one worker came up with a wonderful idea or the answer to a puzzling problem, it wasn't easy for him or her to share it with colleagues. Also, PCs set up as standalone units were very hard to maintain. Each time someone wanted to load a new software package, a utility, or even an update, the process was time consuming.

For all the power that PCs brought business people, few productivity increases could be claimed.

The 1990s: The Network Era

Our current computing era is that of network computing. This is the era of connecting computers together. Network computing conceptually connects all a company's computers, whether they are mainframes, minicomputers, or personal computers.

Network computing also recognizes (and rationalizes) the difference between minicomputers and PCs. There are times when a specially tuned computer should be used for a business application. For example, large databases with customer lists or inventory information are best kept on a specialized computer (or server) that is properly tuned for the task of serving up data.

Technically speaking, network computing lets people share resources on any computer that is hooked up. It also lets them share other devices like printers or extra disk space for storing files. Additionally, it lets computer support people reach out and support any computer on the network.

Network computing lets people build very flexible computer systems. People can plug in or unplug appropriate computing devices—depending on their specific business situation or technology options. Thus, they can take advantage of mainframe computers, minicomputers, or microprocessor-based computers as they need to or want to. Such a simple concept buys a lot of power.

Socially, the network computing era has brought as much change into organizations as all the preceding computing eras combined. Because network computing lets people share information, questions, and answers, they are much more likely to collaborate. Moreover, new applications tools, like electronic mail (e-mail), let people send or receive information regardless of where or when they choose to work.

The 2000s: The People's Era

The next era of computing, according to some experts, will depart from machine dependence and be more focused on individual people. In a people-centered computing environment, there will be many computing devices, such as data servers, networks, desktop computers, mobile computers, and personal communicators.

Ideally, people-centered computing will let people take their connections to data with them—no matter where they travel. Data, information, and messages will be accessed effortlessly.

Already we are seeing the early signs of people-centered computing. In 1992, a consortium of communications and computer vendors introduced the first wireless messaging system in the U.S. The kit, called the Viking Express, combines a palmtop computer, a radio modem, and access capability to millions of electronic mail users. The business person carries a system in a small lightweight leather case that sends and receives messages no matter where she or he is. The user can interactively chat with other users currently on the air. This system fosters a virtual connection between people, allowing them to collaborate even if they aren't in the same city or time zone.

Other services, such as news broadcasts, schedule updates, phone lists, price lists, and other information follow the message capability. Again, they add an ability for information to find a person—instead of a person having to find his or her information.

Major Contributions of the '80s and '90s

From our perspective, the last two decades of computing have been the most exciting. Since the dawn of the personal computer era, and the adoption of low-cost, easy-to-use computers, the number of computer-literate people around the globe has skyrocketed.

As recently as the mid-1970s, there were only about 200,000 computers on the planet. Now there are hundreds of millions—and nearly as many people who know how to use them.

We see the trend toward computer literacy continuing until nearly every adult and child knows something about computing. No one will need a tough specialized education to be a computer user, any more than one needs one to be a telephone user.

In fact, the goal of several computer industry leaders is to make computers as simple to use as a telephone. Intel's energetic president, Andy Grove, once shared this vision during a speech at a major industry trade event. Grove related the dilemma of the telephone companies in the early 1900s. He quipped that, at one time, AT&T conducted research about the requirements to fully exploit long-distance telephone service. The study revealed that AT&T had a huge shortage of telephone operators and couldn't offer long-distance service without a big effort to bring in new people. In fact, it was estimated that, in order to meet the demand by 1950, AT&T would have to hire every man, woman, and child in the United States and train all of them to be telephone operators.

Instead, the decision makers at AT&T decided to throw technology at the problem. They made the process of making a long-distance telephone call easier. Rather than choosing specialized equipment and special connecting cords, they used sophisticated switching equipment that let the connections for long-distance calls come from almost any common telephone. And, in effect, they did turn every man, woman, and child into his or her own telephone operator.

Now, Grove and other computer industry leaders want to do the same thing with personal computers, making them so easy to use that people use them without reservation or a great deal of training.

End-User Computing

The idea of end-user computing is catching on. In the United States during the 1980s, many large companies began the process of training business people to use computers. They developed departments called information centers or end-user computing departments, whose charter was to help business people get their own information.

This practice differed from the older computing model in business, where specially trained information services professionals were creators and disseminators of all information. It also began to spread the responsibility for information out to business people, not solely to computer people.

According to experts, the role of end-user computing will continue to proliferate throughout the 1990s. In fact, Boston University researchers Sproull and Keisler state in their book *Connections* that there will be little concern for non-computer users in business—because everyone will be a computer user.

Not everyone agrees that business people will be universally computer literate. In fact, some industry observers have expressed concern that there will be two classes of white-collar workers—the information haves, and the information have-nots. The differences could well define successful people and successful companies in the future, especially if the predictions of management gurus like Peter Drucker come true and society moves to a knowledge base.

Multiprocessing

Another contribution of the 1980s and 1990s is multiprocessing computing. As a technology, computers based on multiple microprocessors have become very powerful. They stand ready and able to take over the tasks of the large mainframe computer.

Multiprocessing adds another price-versus-performance consideration to traditional ways of looking at computing. From our research, multiprocessing computing almost always delivers more computing power for the dollar than other forms.

Networking

As we mentioned earlier in this chapter, network-based computing is the important model for the 1990s. Not only does networking connect a company's computers together, it connects the company's people too.

Network computing tends to bring in whole new ways to collaborate. Frequently, this is seen in the adoption of special software like e-mail or groupware. Sometimes, it is the simple sharing of documents, like spreadsheets and word processing documents.

We know of one sales department in a major consumer company where employees used their network to share presentations. They set up a shared area of the network and saved all their clever sales presentation materials there. When it came time for someone to create a new sales pitch, there were plenty of samples, examples, and templates to use as starters.

This sharing saved everyone time by removing the need to create new sales presentations from scratch. But, it also lent a great deal of uniformity to sales presentations. The group developed a good style and stuck with it.

Kid-Proof (Executive-Ready) Software

A sometimes overlooked benefit of computing in the post-PC era is the emergence of easy-to-use software. Now that software developers have low-cost, high-powered computers to develop with, they can (and do) feel free to add all kinds of features to make software easy. In fact, we call it kid-proof software. It's easy enough for kids to use.

Software in certain environments, such as Microsoft Windows or the Apple Macintosh, is very visual, easy to learn, and easy to use. The interface (or look) of the software is called a graphical user interface (GUI, pronounced "gooey").

Above all, this software is consistent. So, once you learn a little about its format and how to maneuver with it, it becomes predictable. That's what makes it so easy for children to use.

We think it is also easy enough for non-computer users, even corporate executives, to learn. In fact, we have observed much interest on the part of executive computer novices in this new breed of software.

Client/Server Architecture (the Division of Labor)

A final benefit of newly released technology is the acceptance of *client/server architecture.* Client/server is simply a way of splitting the computing tasks of an application between two or more computers. There are separate roles for the *client* (usually your local PC) and the *server* (usually a specially dedicated computer).

Client/server computing has become a hotly debated topic in the industry over the last few years. Its proponents swear by it, often citing many applications that can be quickly deployed with its tools. Opponents feel that its benefits are overstated.

We side with the proponents of client/server, having seen many applications built in such a flexible manner that, no matter what happens to technology or no matter how the business might change, the application can be modified easily. That's the goal: efficient day-to-day operations and easy modification when necessary.

Multimedia Computing (Voice, Data, Video, Images)

Multimedia computing, that is, the combination of voice, data, or images across the computing wires, got its start in the late 1980s. This type of computing is largely used in educational forums today but may span to other applications as time goes on.

Multimedia computing depends on powerful hardware and sophisticated software to operate well. Many software vendors are beginning work with multimedia kits that store voice and video images in computer files. It is possible to put on a very exciting show simply from the computer screen.

Mobile and Wireless Computing

The 1980s brought the first portable computer. It weighed 30 pounds and looked a lot like a sewing machine. Originally, it had a small green monitor, a couple of diskette drives, and a full-sized keyboard. It required AC power; it didn't run on a battery. It was ruggedly built, and salespeople would routinely demonstrate its fine construction by dropping it on the floor from tabletop height. Most of the time, the computer still worked.

From about 1983 to 1990, mobile computing technology changed so fast it was hard to keep up with. Major product changes occurred almost every six months—breakthroughs in size, weight, screen quality, disk capacity, or other areas.

Today's mobile computers range in size and weight from a unit under one pound that fits in your palm and runs on common AA batteries all the way up to a luggable 20-pound unit with a crisp and sharp color monitor. The choices in between are impressive.

Depending on your business needs, you can find a mobile computer to fit just about all occasions. Most popular today is the so-called notebook, which is about the size of a sheet of paper (8½ by 11 inches), runs on rechargeable batteries, and weighs about six pounds.

We believe advances in mobile computing will continue throughout the 1990s. Computers will continue to get smaller and lighter and more powerful. Functions like wireless communications capabilities and pen input will also help open up the mobile computing options.

Where Is Technology Going?

No matter what angle you look from, technology remains in a constant state of change. We see no end in sight.

Our advice for picking technology is to pick open or popular standards. That applies mostly to hardware but to software as well.

We do not favor any proprietary hardware that will lock in customers. There were too many working computers thrown away in the 1980s because they no longer served a particular purpose and couldn't be used for anything else.

We strongly recommend buying from a reputable hardware vendor. The PC price wars of the early 1990s leveled the pricing field between no-name and name-brand computers. The small difference in price is worth the assurance that your vendor will be there to answer the phone when and if you ever have a question.

On the software side, we believe "popular standards" can be used. For example, while Microsoft Windows is made only by Microsoft, it has become a popular product that is widely supported by vendors in the industry. The same holds true for the network operating system, NetWare, made by Novell.

In our next chapter, we will turn our attention to the convergence of business and technology goals and look for ways to handle the fusion.

Chapter 3

Changes in People

As if it weren't enough to watch the wild gyrations of business and information technology, now we come to the most interesting part of our three-ring circus. It's the people part.

These days, people in business are changing. They are growing more computer literate and more information aware. Also, new work models are emerging, such as cross-functional project teams, self-managing teams, and virtual teams. These have shaken up the status quo. You can't always expect to be physically located with your workgroup members, and sometimes you don't live in the same city, inhabit the same time zone, or even speak the same language. You may not have a boss, or you may find that everyone on your team is your boss. And, sometimes, your closest colleagues don't even work full time for your company—they are rent-an-experts.

Many of the business trends that we talked about in Chapter 1 have had a profound effect on the way people approach their work. Globalization, reduced time to market, employee empowerment, and a myriad of other factors have caused people in business to change the way they approach work.

This chapter examines the people side of computing. We talk about the individual and organizational events taking place that encourage, if not force, people to become more computer literate. We outline a group of six mythical personality types and describe how they interact with technology. Some of these people may look familiar to you. In fact, you may even see yourself in one of the characters.

What Are the People Issues?

People themselves are changing. On balance, they have accepted computers as part of their everyday life, they are getting smarter, and they are more computer literate. In the United States most of us are card-carrying ATM users and perform much of our banking via the automated teller machines. And, with the acceptance of voice mail in many organizations, we've turned our telephone keypads into messaging computers.

Now that PCs or workstations have been in many organizations for a decade, we also find that many business people have grown comfortable with (and sometimes dependent on) computers. We know of managers in some companies who now feel the network-attached PC is as important as the telephone—and who wouldn't think of asking a white-collar professional to work without one. Moreover, the workers agree; many have demanded proper computer connections as a condition of taking jobs. The network-attached PC has become a toolbox of the information worker's trade—much as the carpenter's tool kit with hammer, screwdriver, and nails are to him. As people progress in their ability to use information tools, they're likely to become more demanding. It won't be just a matter of having a computer, it will be a matter of having the "right" computer loaded with the "right" software tools.

Several industry studies point to PC proliferation among white collar workers. One industry estimate showed that 60 percent of the business PCs in the U.S. were connected to networks by the close of 1992. Another study, of 15 large companies, showed a 70 percent network connection rate. Moreover, the study showed that workers felt that having a PC was a "right," not a "privilege."

Hints About the Next Generation of Workers

We believe that, not only is today's business person more computer literate, tomorrow's computer user will be downright demanding. Computer literacy rates will shoot way up in the next ten years as children of baby boomers enter the workforce. Many of the youngsters of the 1980s and 1990s have grown up with a computer in the house. These children, schooled by Sesame Street and entertained by MTV, are very visual and have high expectations of how education should be combined with entertainment.

Since the early to mid 80s, many school systems have been installing and teaching with computers. Children have learned techniques that make accessing data or performing work through a computer quite natural.

As grown-up business workers, the kids of today are going to be much more demanding. They'll expect their computer environment to be tailored just the way they want it—and they won't accept compromise. To illustrate how demanding future corporate workers might become, we share the text of an e-mail message sent by 10-year-old Justy Currid (the author's daughter).

Justy had been given access to a CompuServe account so that she could communicate with Mom when Mom was on business travel. Justy would log on to CompuServe, compose a message, and send it on its way to Mom's wireless RadioMail e-mail service. This method let Justy's message find Mom wherever business travel took her—in airports, taxicabs, business meetings, and so on.

The setup was working fine until Mom let some office colleagues borrow that particular CompuServe account number temporarily. Here's the text of Justy's message concerning sharing her e-mail account:

```
Date: 15 Apr 93 08:56:52 EDT
From: Justy Currid
<55555.1212@CompuServe.COM>
Subject: Yo!
To: Cheryl Currid
<currid@radiomail.net>
Yo Mom,
Will you send me some messages & get me an account all
of my own? Because when I went to check (MY) account I
found messages from people I have never heard of. And having to
keep up with strangers is out of this world!
JUSTY
P.S. you know what I mean, don't you?
```

Besides getting a chuckle out of Justy's insistence about getting her own "private" e-mail account, we think this gives some insight into the expectations of future corporate workers. They are likely to be much more insistent and aware of what they can get. The proper computer configuration will be a right, not a privilege.

Demands of a Knowledge-Based Society

As we move toward a more knowledge-based society, many individuals are looking for ways to get smart. They need to know as much as they can about their jobs, their industry, their competitors, and their options.

Computer literacy is already beginning to separate the information "haves" and "have-nots." In some companies, an edge in personal computer literacy can be the deciding factor in getting the next job (or getting a job in the first place). It's easy to see why. Let's say an employer has the option of hiring two otherwise equal candidates, but one has a high degree of self-sufficiency and can get to the root of business issues and problems. It stands to reason that this person will be a better contributor to the effectiveness of the company than the other, who might have a similar academic background and as many years of work experience—but less personal ability to dig into the information base of the job.

Today, however, computer literacy is not universal, even among white-collar workers. There are some personality types that have a tough time getting the picture. Whether it's Teddy Too-Cool-for-School, or Kevin Can't-Learn, there is a constituency of people who have not found computer literacy normal or natural.

Several years ago, we put together a spectrum of computer users as a way to define literacy rates. We divided groups based on their personal abilities to work with computers. Our breakout of personalities included six classes of white-collar workers. We identified them by the character names:

Kevin Can't-Learn
Teddy Too-Cool-for-School
Wanda Want-to-Be
Joe Average
Frank Flash
Steve Splash

Figure 1 shows a pictorial representation of the relative numbers of each type of user you might expect to find in a progressive company. Clearly, most users fall into the middle ground, possessing an average or mediocre degree of computer literacy. We think that this graph will change over time, moving to higher levels as more and more people climb the computer literacy ladder.

We offer some specific information about these personalities, as well as an action plan for advancing some of them to higher levels of literacy.

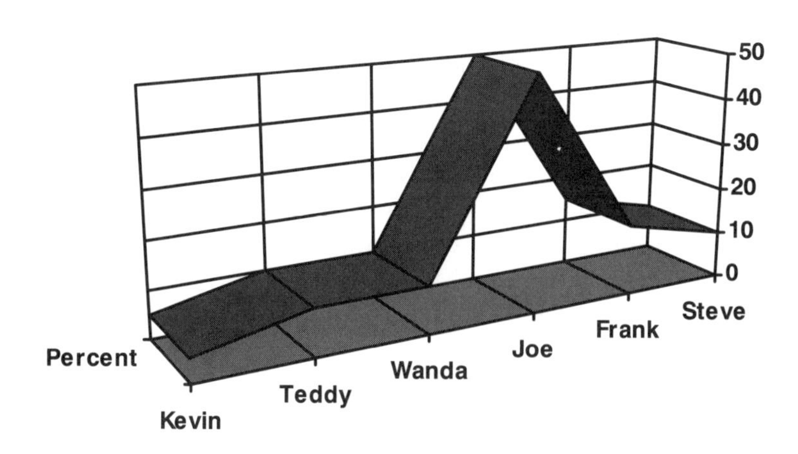

FIGURE 1 Computer users' literacy spectrum

Don't be surprised if you see yourself in one (or more) of these categories. (Also, please note our cutesy alliterative names were not intended to represent gender in any way—they're just cute names that people could remember.)

Kevin Can't-Learn

Kevin Can't-Learn is a personality characterized by just not connecting with computers. Kevin and his cousin, Catherine, continue to flunk even remedial courses of Computer 101. They are keyboard klutzes and, try as they may, they simply can't remember what buttons to push from session to session. Their employer has spent countless hours trying to train them—in spreadsheet programs, word processors, and databases. But, alas, they just don't get it. Sometimes you wonder if it wouldn't be more merciful just to take Kevin and the clan out to the side of the building and hand them pink slips.

While we might make jokes about the Kevin personality type, these people really do exist. Unfortunately, when they move into certain knowledge-based white-collar jobs, they become a burden to everyone. They drain training resources; they get to know the company's computer help desk staff on a first-name basis; they need other people to pull information and do work that they should be able to do for themselves. They also tend to drag down the productivity of teams. For example, if Kevin can't figure out how to send and receive information on the company e-mail system, then he will forever be kept from certain team-related activities and decisions.

We haven't given up hope on Kevin. We believe that the Kevins of the business world can be brought up to some speed, but *only* if they take personal responsibility for their own computer literacy. That means taking the extra effort to become computer literate. Generally, slow learners cannot learn computers and do their full-time jobs too.

An Action Plan for Kevin

If you recognize Kevin in yourself (or in an associate), our prescription for change includes:

- Buy a personal computer for home. (We recommend one with Microsoft Windows pre-installed.)
- Take a local community college or trade school course on Windows software and learn the basics.
- Learn how to type. Often the problem of Kevin-types is that they are keyboard klutzes who never learned to type. They have problems mastering two skills at once—the mechanical skill of typing and the thinking skill of the computer application. By learning to type (just a little), you remove one of these barriers.
- Subscribe to an online information service, such as CompuServe or Prodigy, and vow to use it for something three times a week. These services often have forums, or electronic bulletin boards, on a variety of subjects. Members exchange information and tips on hobbies and other interests on a regular basis. These services also offer a host of areas providing current stock market information, news, weather, and sports.
- Buy personal finance software (such as Quicken for checkbooks) and use it religiously.
- Join a user group or find a "computer buddy" to talk to. It is helpful for the learning process if you exchange information with someone on a regular basis.
- Buy office productivity software (a spreadsheet, a word processor, and maybe a graphics package or filing package). Bring some work home and set up something. People find it helpful in learning these packages to apply a real business problem to them. Consider using a

spreadsheet for some calculation or a word processor to compose a letter, memo, or report.

- Enroll in classes for office productivity software (one class at a time). Master the package before you move on.

The purpose of the program is to increase your frequency and use of computers for varied tasks. This process builds familiarity and habits, which will turn into skills. We also recommend that Kevin-types first focus on "using" computer applications that somebody else built, not trying to get into the learning and building process all at one time.

Without action, we think, Kevin-types are going to get pushed around and out of corporations until they find a suitably undemanding position.

Teddy Too-Cool-for-School

Our mythical Teddy personality is usually a corporate or business supervisor, manager, or executive. This mid-to-upper-level individual often is in a position to shape decisions for the organization, so what he or she knows is important.

Often, he and his female counterpart Theresa ascended the rungs of their career ladder before computer literacy was compulsory. Today, they have a staff of people, such as secretaries, assistants, or researchers, to do the information gathering for them. Sometimes this staff acts as a filter or shield and keeps certain information away from the boss.

The problem with Teddy-types is, although they may be smart in their chosen professions, they frequently lack mechanical skills (they never learned how to type very well) and procedural skills (a programming language would send them to the next planet). To compound the problem, Teddy-types often need their information on demand, in cluttered and unstructured ways. Depending on the situation, they may need information from many different sources outside the company's own data banks. High-level or strategic decisions often require information from far-off places, like new services, outside agents or brokers, or reports of competitors' activities. Depending on the specific problem, it isn't always easy to say what will be needed and when.

In some ways Teddy-types can be held hostage by these information providers. Since they have no personal computer skills, they can't always challenge what they are told. This makes Teddy-types vulnerable to making big decisions (and big mistakes) with misinformation. Moreover, Teddy is placed at a disadvantage when an agile, self-propelled colleague (or competitor) comes along who can make faster or better business decisions.

A company with a senior management team of too many Teddy-types is at a severe disadvantage in the 1990s and could easily goof enough to go out of business in the 2000s.

An Action Plan for Teddy

If you are (or know) a Teddy-type, know that there is a definite cure for computer illiteracy. After all, these people are smart people in their chosen profession—they just didn't get around to learning about computers. Our prescription includes:

- Get a portable (notebook) computer and carry it everywhere. Take it on business trips, home for evenings and weekends, even to the golf course. (Again, we recommend equipping the computer with the Microsoft Windows operating system and Windows-based application software.)
- Hire a computer coach or find a set of video training courses to learn the basics of PCs and Windows software. This elementary learning is important. It makes Teddy-types not only more comfortable with computers but more confident too. (It is also important that basic training be handled discretely. Most Teddy-types don't want to admit what they don't know. They need to have someone or something explain the very basics of computers.)
- Make e-mail the first application to learn. E-mail is the single most used application for executives. It frees them from having to track down people in the organization—and it also lets people get to them. We'll talk more about the power of e-mail in Chapters 14 and 15.
- Learn how to type. Since e-mail will likely become one of the executive's most important tools, and since handwriting and voice recognition are a far cry from developed technologies, we don't think executives can escape learning some keyboard skills.
- Buy a personal information manager (PIM). This is a special application that collects information like name and address lists, notes, thoughts, and sometimes appointments and other dates. Make sure the PIM software works with the software at the office and that subordinates can update the data. It makes no sense for a corporate executive to take the time to maintain a personal address book. This is a non-cerebral task that can easily be delegated to a staff assistant. Just

make sure the PIM software allows an easy way to transfer the updated records to the executive's computer.

- Hook up an executive information system, including some form of easy-to-use query tool. Most organizations have valuable vaults of data that are too hard to get to. With a good database program and query tool, the executive can start getting his or her own information—and do so in ways that let answers prompt further questions.
- If time permits, consider subscribing to an online information service, such as CompuServe or Prodigy, to keep up with business and travel activities. Also, be sure to hook up a news feed that automatically clips pertinent events on the company, its competitors, and the industry. Getting timely news to Teddy-types is critical.
- Buy office productivity software (a spreadsheet, a word processor, and maybe a graphics package or filing package) last. Executives frequently don't have time to learn these packages. They can master them later, as time and familiarity permit.

The purpose of Teddy's program is to highlight the opportunities for quick-hit productivity. Frequently, these opportunities center on time, decision support, and communications management. Without action, we think, Teddy-types will not just hurt themselves, they will hurt their organizations.

Wanda Want-to-Be

The blossoming of Wanda and Walter Want-to-Be illustrates an interesting phenomenon of the personal computer era. Unlike Kevin, these people really can connect with computers. They discover some form of latent technical talent that they did not find during their years of formal education. As a result, they sometimes land in lower-paying jobs with responsibility for gathering but not analyzing information. Their jobs don't require that they use all of their brain power.

Even though Wanda-types have the brain power, they just didn't get involved with information technology beyond a calculator during their early, formative learning years. It might have been for economic or social reasons—maybe they didn't have the interest or funding to seek higher education or technical training. Or perhaps they weren't mature enough as youths to recognize their calling to computers. Whatever the reason, they simply didn't develop their brain power.

Wanda-type employees can be found working as secretaries, statistical clerks, administrative assistants, or other white-collar helpers. Frequently, they feel intellectually unfulfilled by jobs that lack the challenge to create or innovate. Their current jobs are often pretty static and routine.

Then comes the computer—and the unleashing of previously locked-up talent. For whatever reason, Wanda and Walter meet the computer, and nobody can seem to get them off. All of a sudden, these people get connected, and all kinds of innovation and creativity start to pour out. They begin to take projects on their own initiative, like developing new departmental tracking forms, redesigning the division's status reports, or some other local productivity project.

Many companies are surprised to find Wanda-types and don't know what to do with them when they emerge. Human resource departments are at a loss because these people don't fit neatly into a category.

An Action Plan for Wanda

We think Wanda-types can be a powerful asset to a company, but only if given additional education and more latitude in their jobs. All too often, Wanda and Walter turn into dilettantes. While they have plenty of capability, their brilliance can be frittered away on less meaningful tasks. Without a constructive path, Wanda-types will end up writing meaningless macros and merge files that nobody really needs.

Human resource departments may overlook corporate Wanda-types because they do not possess standard requirements for existing job descriptions. This can cause large corporations to under-use an important asset.

Someone, either Wanda herself or her boss, has to decide whether or not to develop her skills and brains. If you know of (or are) a Wanda, we offer two plans of action for either retarding or encouraging Wanda's literacy:

To Discourage Wanda's Strategy...

This action plan is limited to containing Wanda-types to the specific job they were hired for. (We don't like this approach because we believe that all business people should be challenged to be the best they can be. But, recognizing that there are business situations where career progression isn't possible, we offer the following suggestions):

- Hold Wanda to the specific tasks of the job. The Wanda-type personality will wander into any kind of additional activity, project, or mission—even outside the normal business hours. If Wanda is to be contained, don't allow special assignments or extra work. Get Wanda a season ticket to the movies or a bowling alley.

- Do not encourage any additional computer training. The more Wanda learns, the more Wanda will want to do. The training will also take time away from the job Wanda was hired for.
- Keep Wanda away from new technology.

To Encourage Wanda's Strategy...

On the other hand, if Wanda is to become truly productive, additional education is important. In some ways, the prescription for Kevin (except the typing tutorial) will jump-start Wanda-types and fill some of the holes in their technology training. These tasks include:

- Change Wanda's job (a little at a time). Give Wanda additional responsibility that includes work on a computer. In the beginning, get professional help to make sure Wanda's creations are correct. We know of too many cases where Wanda-types went off and created wonderful but useless spreadsheet macros. They were wastes of time and budding talent.
- Buy Wanda a personal computer for home. (Again, we recommend one with Microsoft Windows pre-installed.)
- Offer to reimburse Wanda for enrolling in a local community college or trade school course on Windows software, where she can learn the basics. Then, suggest that she take courses on database basics and application development tools. By hook or by crook, Wanda-types will end up developing some kind of application for the company. Make sure that development is done using the proper tools in the correct way. We know of too many cases when Wanda-types built disastrous applications with word-processing macros that should have been done in a database. Had Wanda known the database tool existed, and had she the skills to use it, that effort would have flown and not flopped.
- Subscribe to an online information service such as CompuServe or Prodigy, and let Wanda wander. The forum portions of these services will let Wanda meet other electronic friends and share ideas.
- Buy Wanda personal finance software (such as Quicken for checkbooks) and suggest that she use it religiously.
- Find Wanda a computer buddy who is a professional computer person. Often, Wanda lacks the disciplined education that information

services (IS) staff people have. It's a good idea to team up Wanda and a professional IS person so that they can share techniques and experiences.

- Direct Wanda to user groups and special interest groups. A lot of learning can be accomplished in these sessions.
- Other suggestions, such as buying office productivity software (spreadsheet, word processor, and maybe a graphics package or filing package) are a given. Chances are, Wanda is already an expert in one or more of these applications.
- Consider letting Wanda train others. Often Wanda-types are brimming with excitement and enthusiasm about computer technology. Many companies have successfully capitalized on that energy by letting Wanda train other employees.

The purpose of the program is to fill in any gaps in knowledge and then get Wanda-types advanced training on tools. Once the Wanda-type has a full toolbox, he or she will soon become a master at using the right tool for the job.

Joe Average

Joe and Jane Average may represent as many as 50 percent of the white-collar workers in today's progressive organizations. These people are in occupations that require them to have information and knowledge in order to do their jobs well. They can be accountants, marketing specialists, human resource professionals, or other white-collar workers.

Career-wise, they are often in competition for the next job and can best show their potential by showing how smart they are and how much they know. In fact, the more they can show what they know, the better.

In short, Joe and Jane are already computer literate. Long ago, they found the on-off switches for their PC; they know where they last stored their files and how to get them back; they can create their own spreadsheets and even get most of the calculations right. These people are computer literate and functional. They are not, however, computer gurus and never want to be.

Many companies make the mistake of thinking that Joe and Jane don't need any computer support or training. Executives figure that, because Joe and Jane can calculate a column of numbers on a spreadsheet or issue a report using a word-processing software package, no further training is necessary.

This state of affairs can be a problem. The problem stems from the fact that, although Joe and Jane do know how to use computers, they do not know how to use them efficiently.

They tend to work hard but not smart, to use the wrong tools, to kludge up their efforts, and sometimes to make pretty big goofs. Often, Joe and Jane learned only the rudimentary features of the software packages they use—they never got to the advanced or time-saving methods. This gives them less-than-efficient habits. For example, most word-processing packages include powerful list-merge features so that people can send out apparently personalized letters to a number of people with very little effort. But if Joe never learned how to use this feature and has to send out 50 letters, chances are he'll spend five to ten times as long as necessary individually addressing each letter. He has the tool, but he doesn't know how to use it.

Another common problem for the "Averages" is that they use the wrong tool for the task at hand. We have seen, all too many times, accountants typing status reports into the cells of a spreadsheet (because they never learned how to use a word processor). Outside, their secretaries are committing a similar sin by calculating an expense report using tables in a word processor. In both cases, they are using the wrong tool for the job and paying the consequences of lower efficiency.

An Action Plan for Joe

We think Joe and Jane Average are partially the cause of today's business productivity paradox: the complaint that companies have spent billions of dollars on technology yet see very little increase in productivity. Joe and Jane Average need to accurately assess their personal skills, their jobs, and their needs to learn new tools.

Our suggestions include:

- Buy a personal computer for home or a notebook computer that can be taken along if business travel is a part of the job. (Again, we recommend one with Microsoft Windows pre-installed.)
- Take advanced training courses in application software that is used on a frequent basis, such as spreadsheets for accountants, graphics software for business presenters, or advanced word-processing techniques for anyone who needs to create letters or complex reports.
- Learn a database. Many Joe-types stop their education in computer software just short of where they need to. They learn spreadsheets, word processing, e-mail, and graphics, but not a database. Yet, for many knowledge-based jobs, it is the proper understanding and

manipulation of data that distinguish great analysis from mediocre work. Database software, available for PCs, is a must-learn for anyone who has to maintain or analyze information. Even if Joe never has to develop a database personally, he should understand the principles behind databases.

- Subscribe to an online information service, such as CompuServe or Prodigy. The forum portions of these services will let Joe meet other electronic friends and share ideas.
- Purchase a CD-ROM drive and a software package that includes an encyclopedia and writing tools (such as a style guide, a dictionary, a thesaurus, and other manuals). These packages open up a new world of learning to Joe-types. There is a wealth of information available on CDs, starting with the information contained in packages. New CDs are introduced each month with back issues of popular magazines and research materials. Starting off with the basics should help open Joe's eyes to what's available.
- Buy personal finance software (such as Quicken for checkbooks, Turbo Tax for taxes) and use it religiously.
- Get a computer buddy who is a professional computer person or very competent business user. Joe-types will pick up new techniques a little at a time and begin to apply them. It's very helpful for Joe to learn from other advanced business users (like Frank Flash and Steve Splash, described next). In this way, Joe will learn not only more about computers but also more about business situations that can be addressed with computer technology.
- Join user groups and special interest groups. A lot of learning can be accomplished in these sessions.

Much as in our prescription for Wanda, the purpose of this program is to fill in any gaps in knowledge and then to get Joe-types advanced training on tools. Once the toolbox is full and the training is enhanced, Joe and Jane will become much more productive. In fact, we believe, these people can quickly advance to our next categories when they are pointed in the right direction and given a little incentive to go forward.

Frank Flash

Further up the computer competency scale is Frank Flash. He and his counterpart Frieda are fast becoming advanced users of technology. They don't claim to be experts, but they've mastered the information tools of their trade and are beginning to show great results. Frank-types are convinced that all the information they need is available, but they don't always know how to get to it.

Frank and Frieda are energized. They are willing to learn whatever they need to. They aren't afraid to dig into software packages and spend a little extra time to find the best approaches and techniques. Once they do learn something, they get a lock on how to produce fantastic results.

Frank-types are often found in business as middle- to upper-level specialists or managers. They are very proficient at their chosen occupations and frequently stand out from their peers. They are no-nonsense professionals who use computers as tools to make decisions. In large corporations, Frank-types can drive their information services staff crazy. They don't take time to budget money for their own computers but find a way to talk somebody out of whatever is needed. Of course, nobody likes to say "No" to Frank-types, because they always produce fabulous results that show up right on the bottom line.

A good example of a Frank-type is the logistics specialist of one organization who spent the summer ripping through a software package that reassigned customers to manufacturing plants. The analysis was complex and considered several factors: purchase volume, frequency of purchase, location, and profitability. Relying on a series of elaborate formulas, Frank was able to save his company several million dollars a year through manufacturing and distribution efficiencies. It turned out that the old simple model, shipping products to customers closest to a given plant, was inefficient and often resulted in out-of-stocks, missed shipments, unprofitable orders, and poor customer service. It took a lot of effort—weeks of using borrowed computers, countless hours of programmer support time to pull historical data from the company's mainframe computer, and seemingly endless use of networked printers—but the results more than paid for the effort.

We think companies should locate and encourage their Frank-types. These people will take computer technology and turn it into action.

An Action Plan for Frank

We are almost tempted to give Frank a blank checkbook or endless line of credit at the local computer store. Since Frank-types tend to bring such good results to the bottom line, it might be worth the risk.

- Recognizing that most business people will stop short of a potentially bottomless investment, we suggest that you encourage your Frank-types to take these steps:

- Buy several personal computers for home, office, and travel. Frank-types should never be far from a keyboard. With the low cost of PCs, it makes sense for Frank to have one wherever and whenever he needs one.
- Take advanced training courses in database and analysis software packages. Frank-types already know the basics, but they don't always have the advanced software packages down pat. Where classroom-style training isn't possible, consider hiring Frank a computer coach who will work one-on-one with Frank. Generally a professional computer person will be needed to help Frank with high-end programming tasks. Frank doesn't have to learn a programming language, but he will probably end up learning advanced database extraction tools.
- Make sure Frank-types have access to whatever online information services they need. Services such as CompuServe, Dialog, Nexis, or others with large information bases are necessary for outside research.
- Purchase a CD-ROM drive and software packages that give Frank-types access to outside information. This step should supplement subscriptions to online services. CD-ROMs are generally lower in cost than online services.
- Join user groups and special interest groups for the newsletters. Chances are Frank-types won't have the times to attend meetings, but they will benefit from reading newsletters from user groups.

Overall, Frank-types are the business computer users of the future. They are quick, competent, and nearly self-propelled. They represent an important competitive asset for companies as organizations move into what Peter Drucker calls the "knowledge-based society."

Steve Splash

The super-duper knowledge engineers are summed up in the personality of Steve Splash. More focused than Frank, Steve and his colleague Sue Splash will be the people to see when anyone needs to know anything. They know where to go to get the facts and how to make instant sense of them. Steve and Sue know all the things Frank is trying to learn, like where all the corporate data reside and where all the public data are. They know how and where to tap the reservoirs of data, what's important, and how to pull it all together.

When it comes to solving corporate information issues, Steve-types can drill down to core facts in a nanosecond. For example, we know of one company that was losing sales on a previously popular consumer product. Everyone involved was scratching their heads: advertising looked good, distribution hadn't changed, and no new direct competitors had entered the market. The problem was turned over to a Steve-type and solved within a week. By gathering up all forms of data, Steve found that a new product introduced by another firm was taking away sales. Even though the company didn't feel that the new product was a competitor, the consumer felt differently. Through an exhaustive search of consumer sales data, Steve was able to find an almost direct correlation in lost sales with the new product's gain. As a result, the company was able to target and focus its efforts and bring back consumers. Had it not been for Steve's efforts, the company would have never seen that new product as a threat.

Steve's only shortcoming is not understanding the capabilities of other business people. Steve-types simply don't understand that others might not be so smart. As a result, Steve sometimes creates models, forms, templates, macros, or scripts that others cannot easily follow. This can be a problem if Steve lands in a job where he is creating work not only for his own analysis, but for others. He'll drive Kevin and Teddy to turn off their computers; Wanda will be frustrated; Joe will most likely give up; and Frank will be flaming mad because he can't follow. Steve-types need someone around who can translate their ideas into easy-to-use computer models.

That limitation aside, Steve-types should be given whatever they require to get to the information they need. They are truly super in the way they can use information technology—and they should be granted almost any and all wishes.

An Action Plan for Steve

Just as with Frank, we are almost tempted to give Steve a blank checkbook and endless line of credit at the local computer store. Since Steve-types can do just about anything with computers, there is great temptation to let them go without question.

Instead, we recommend the following loosely defined path for Steve:

- Like for Frank, we think it is justifiable to buy several personal computers for Steve-types. To buy one for home, one for the office, and one for travel makes sense, or get one super-duper unit that can be taken wherever Steve goes. We also think Steve should get the latest and greatest technology. In a large company, this demand is easy to accommodate. Steve should get the newest computer, and his old units should be passed down to other computer users.

- Encourage Steve to take periodic advanced training courses in new software. A common problem of Steve-types is that they become so entrenched with their tool set, they sometimes don't look to newer and more sophisticated software. They tend to immerse themselves in software and procedures that sometimes go out of date, This a mistake. Steve-types should periodically come up for air and look at new tool offerings.
- Make sure Steve-types have access to whatever online information services they need. Services such as CompuServe, Dialog, Nexis, or others with large information bases are necessary for outside research.
- Make sure Steve has all the CD-ROM based information packages that are necessary for the job. These should supplement subscriptions to online services and save money from searching more expensive online services.
- Get Steve a computer coach. Although Steve-types don't think they need any training, they sometimes do. A computer professional or another Steve-type is helpful to make sure Steve isn't becoming out of date by using tools and techniques that he has grown accustomed to.

Overall, Steve-types can become the most productive asset to business both today and tomorrow. Like Frank-types, they are quick, competent, and nearly self-propelled—only more so. They are a competitive asset.

Today's Trends

The trend in business today is for people to become more computer literate and more capable of handling their own information needs. This doesn't mean that professional computer people will go the way of the dinosaur, but it does mean that, along with individuals, jobs will change. We'll talk more about some of the changes occurring now and in the near future in both Chapters 5 and 7.

In the next chapter, we'll focus on getting it all together—the combination of business, technology, and people.

Chapter 4

The Convergence of Business, Technology, and People

> Close your eyes. Imagine a company where the people, the process, and the technology are all working together for the same objective. Imagine that the technology is appropriately applied, that the quality is built into the processes, which are constantly being refined for better service and speed-to-market time. Imagine, too, that in this company the people have the freedom to express themselves and to make decisions to support a specified goal. Sound Good? Now—imagine this company as your competitor. (Rod Canion, Chairman, Insource Management Group)

To get it all together. That's the goal—and you can be sure that some companies will figure out how to do just that. Let's hope yours is among them, and your competitors aren't.

In the last three chapters, we talked about three different variables: business, technology, and people. We used the metaphor of a three-ring circus—and someone (you) as the ringmaster. A difficult job at best.

Now, we are going to change the image from a circus to a real business. We want to envision an *empowered enterprise*, much like the one Rod Canion speaks of in the quotation that opens this chapter. We think the ideal is an organization that can adapt to its ever-changing business environment. It takes the talents of its people and enhances them with information technology. It is faster, leaner, and meaner than its competitors. It can initiate, respond, or react—or simply do what it takes to stay on top.

So, how does a company achieve such a miracle? We believe businesses can go a long way toward perfection by carefully considering, and then mastering, five simple steps. They are:

1. Cultivate a computer-literate workforce.
2. Empower business people with information and authority.
3. Reduce development time of information applications.
4. Regularly reengineer business processes, and then reinforce them with technology.
5. Consider information as an asset.

We are not saying that these five steps provide a complete cure-all for what ails any and every corporation; but we are saying that these steps will go a long way toward making an organization more flexible and capable of handling change.

In this chapter, we'll examine each step in detail. But, be warned, simply reading about or discussing these steps won't provide any answers—action must follow.

Cultivate a Computer-Literate Workforce

First and foremost, organizations must cultivate a computer-literate workforce. As we saw in Chapter 3, not everyone is as efficient a computer user as he or she might be. Sure, the next generation of workers will undoubtedly come already trained with excellent computer skills, but most companies can't wait until the year 2010.

Benefits

We believe organizations must do what it takes to raise the level of computer literacy. The benefits should be clear: the organization gains a smart, quick, and *empowerable* workforce.

Smart Workers

Most people understand the term "smart." Employees are smart when they know the tools of their trade and can quickly get the right answers to questions or challenges. The questions might range from marketing to manufacturing, from competition to cooperation, or anything in between.

Quick Workers

Most companies need speed, whether in the form of quicker product delivery, or in that of quicker market analysis. Organizational workers no longer have the luxury to take time to make decisions.

Empowerable Workers

Empowerable workers are workers who are ready to accept new tools and technologies that will help them do their jobs. As we have said, companies cannot delegate decision making to dummies. There must be an in-shape team of people who are smart and quick before empowerment works.

How to Cultivate Computer Literacy

Organizations should make a concerted effort to assess the computer literacy of their workers. They can make this assessment through a formal study, through self analysis, or informally.

In a formal study, the company management generally retains an outside firm to conduct a survey of computer literacy. These surveys are usually self-reported answers to a prepared questionnaire. The questionnaire is evaluated, and the consultant makes some recommendations.

In a self-analysis, the company IS department usually takes the place of an outside consultant but performs the same basic task. It asks computer users to rate themselves on literacy.

Some organizations take an informal approach by having managers discuss the computer training needs of each department. This approach is subjective and most often leads to misunderstandings. It is likely to perpetuate the "Joe Average" computer user's condition, because it assumes that people know more than they do.

Reading the Tea Leaves

Once computer literacy has been assessed, it is important to follow up with action. In most cases, a survey of computer users will reveal that additional training is needed. In fact, we have yet to see an organization where no additional training is needed.

Train Them

Generally there are two types of training required to cultivate a more computer-savvy group. They are *mechanical training* and *job-related training.*

Mechanical computer training is simply teaching people the mechanics of the software they use. It consists of showing them what keys to push to get something to happen. For example, this type of training will cover the basic commands of a word-processing program—how to create, edit, save, print, or delete documents.

Mechanical computer training is easily accomplished through a variety of sources including: in-house computer training sessions, outside training through local schools or community colleges, instructional video tapes, books, or users' manuals. To suit the needs of a large workforce, it may be necessary to use several of these sources.

Job-related training is a little more focused—and potentially more powerful. This type of training focuses on how to apply a computer tool to a job. It is not how to calculate the sum on a column of numbers; instead, it is how to build a better analysis form with a spreadsheet. We believe that job-related training must be customized for each job or process and can best be applied by someone who does the job every day.

Make Computer Literacy Cool

Another way to cultivate computer literacy is to make it "cool." At one time, many professional white-collar workers shunned computer work as clerk work. They felt that computers were for lower-level people in the organization, certainly not for managers and high-end specialists.

These days, that line of thinking simply doesn't work. To change it however, takes a concerted effort. One company we know made the change by appointing a departmental champion, who was tasked with helping each of his coworkers become more computer literate. He didn't necessarily train them personally; he would just shame them into using the computer rather than calculators or paper notepads for their work.

Another company helped the process along by assigning temporary "evangelists" to workgroups. The evangelists were actually information services staff members who were tasked with the job of working individually with people to help them change work habits. These IS staffers helped out by spending time observing work habits and then working with people to find more efficient work patterns (by using their computers).

Over time, this program helped convert a number of the hard-core illiterates to using the tools that were at their fingertips. It promoted the use of many office productivity tools, like e-mail, electronic schedulers, spreadsheets, and database query tools. In some cases, it uncovered a need for new applications to be developed by the professional IS staff.

Hire Them

Some companies are already beginning to screen new hires for computer literacy. Concerned about the cost of training their employees, they have worked to ensure that each new employee is already well grounded in basic computer skills.

While computer literacy is difficult to test for, it is well worth the effort to uncover. After all, if an organization is composed of highly computer-literate workers, someone who lacks these skills would certainly be out of place.

Empower Business People with Information and Authority

The word "empowerment" has been established in our business vocabulary for years. It suggests that all decisions cannot realistically be made on the upper floors of corporate ivory towers. Instead, the best decisions and actions can (or should) be made by people on the front lines. These people, however, must first be smart enough to make good decisions. Accordingly, front-line people must be armed with the right information, in the right format, at the right time.

It all sounds so simple. In reality, however, giving people the right information, in the right format, at the right time is not so easy. Information systems built from 10 to 20 years ago did not embody such goals. Instead, these vintage systems, known as "legacy information systems," were intended to optimize computer resources, not people resources. Back when they were new, after all, computer time was expensive, but people time was not.

The notion of empowered people requires that everyone recheck information delivery to make sure it is timely, accurate, and complete. If it fails such a test, it should be reconstructed so that it does empower people.

Reduce Development Time of Information Applications

Information technology has played a contradictory role in the adoption of change. The indefinite extension of the backlog in mainframe applications development provides the incriminating evidence. Instead of speeding the flow of information, mainframe applications took months and often years to develop. Consequently, operating departments had to compete for data processing resources to develop their applications.

When minicomputers were introduced, packaged software became abundant and sometimes shortened the waiting time for applications. With the adoption of personal computer technology, operating departments found ways to develop their own applications, using spreadsheets and database management systems. They could generate reports in the format they needed and even move data to a graphics package for presentations to managers and clients. Companies also found that they could connect the PCs to a mainframe, download data into spreadsheets, and analyze the information without waiting for a programmer to write new software. Individual productivity often got better, but organizational productivity sometimes didn't.

At that point, companies realized that they could save both time and money by moving applications off mainframes onto LANs. Because LANs allow the use of low-cost, quick development tools, they make for powerful application development environments.

Reducing development time of applications often involves some fundamental changes in a company's core computing platforms and development strategies. It requires that companies install advanced computer hardware, software, networks, and methodologies.

This requirement generally means buying computer hardware with "cheap MIPS." (A MIPS is a standard unit of computing power, which stands for Million Instructions Per Second.) That is, a company buys computer hardware based on low-cost but powerful microprocessor chips, such as PCs and PC-based servers. Computers based on the Intel microprocessor platform are generally the most cost-competitive.

Software tools must be changed out too. Organizations must move from older computer languages, such as COBOL, into advanced development environments. These environments, such as Visual BASIC, PowerBuilder, SQL Windows, and others all provide the developer with easy-to-use tools that can cut the time it takes to create many business applications by as much as 75 percent.

Finally, development methodologies must also be changed. For all but the most difficult and complex business applications, the best approach is prototyping.

In prototyping, the business user and developer work together to sketch out a functioning model of an application. When the model is complete, it is actually placed into business use (as a test). It is refined on a continual basis until it meets business needs. This type of approach actually works faster and delivers better results than older methodologies that asked users to envision and sign off on their needs before they ever saw the end result.

Regularly Reengineer Business Processes

Rather than continue to cast information systems in the likeness of ingrained but inefficient processes, analysts are beginning to ask probing questions. Challenged to reengineer the business process, more and more people who build new computer systems are starting off with a blank sheet.

This means asking a whole lot of "why" questions. Why do our accounting clerks carefully circle selected pieces of invoice information, run manual batch totals, and then send the work to data entry? Why do they fill out paper reports and then forward them to a regional clerk for inputting into a computer? Why don't they have the information downloaded to their desktop machines and make the corrections themselves?

We applaud the exercise. Rather than assume that everyday work flows are correct, you must challenge the status quo. But we also think it takes some really innovative thinking to make the questioning process work. Few business people have been schooled in the fine art of asking why.

So, there are new skills to learn. The skills start with looking at business processes and asking why. We can't accept a 26-step work flow as being right, efficient, or even morally correct. Instead, we need to focus on why, and then put on our thinking caps to see if we can find a better or faster way.

In one organization, for example, it could take upwards of a week to process a travel request form or a purchase requisition for supplies. People had to jump through hoops and loops to do it.

Consider this official 15-step procedure that a large company had been following until recently:

Step 1: A business person handwrites a travel authorization request (TAR) to take a trip to deliver a paper at a conference.

Step 2: The business person handwrites a memo with the justification for the trip and hands the TAR and memo off to a secretary.

Step 3: The secretary calls the company's travel agency for prices of airline travel and calls the hotel that the business person prefers to determine room prices.

Step 4. The secretary edits the memo for spelling and grammar.

Step 5: The secretary types the TAR multicopy form and the justification memo and returns them to the business person.

Step 6: The business person reads and checks the TAR and memo, signs both, and hands them back to the secretary.

Step 7: The secretary walks the TAR and the memo down the hall to the department manager's secretary.

Step 8: The department manager's secretary checks the TAR and gives it to the department manager to sign.

Step 9: The department manager reviews the TAR and memo, signs both documents, and returns them to the department secretary.

Step 10: The department secretary removes the back copy of the TAR, walks to the copy machine, and makes a copy of the memo. She puts the copies in her file cabinet and calls the business person's secretary.

Step 11: The business person's secretary picks up the original copies of the TAR and memo and takes the documents to the copy machine, where she makes three copies. One is for the business person, a second is for her permanent file, and a third is for her daily tickler file.

Step 12: The secretary sends the originals of the TAR and memo to the accounting department.

Step 13: On the day before travel, the secretary takes the TAR from the tickler file and walks to accounting to pick up the business person's travel advance and airline tickets.

Step 14: The secretary gives the tickets and check to the business person, who signs the copy of the TAR and returns it to the secretary.

Step 15: The secretary files the TAR until the business person returns.

(Elapsed time: three days to one week.)

Now, consider the way things are done today with a reengineered process and a little help from the technology of electronic mail. Today, the process is down to four steps.

Step 1: The business person drafts and spell-checks his or her own TAR and memo in a word-processing program, using a template for the TAR form.

Step 2: The business person sends the electronic TAR and memo attached to an e-mail message to the department manager down the hall.

Step 3: The department manager down the hall approves the TAR electronically and forwards it to the accounting department and to his secretary.

Step 4: The department manager's secretary makes reservations for the business person.

Elapsed time: 3 minutes to an hour.

Today's system (even if it does require a PC and LAN connection on everybody's desk) costs less than the old-fashioned approach, if you consider the cost of labor.

If reengineering works so well in this example, can you imagine what the process could do in other cases? It doesn't take long for the benefits of changing business processes to start adding up.

In fact, we believe, business processes should be measured and evaluated on a regular basis. Critical processes (such as those involved in making or selling core products or services) might well be reviewed every two to three years, while non-critical processes (such as administrative procedures) should be reviewed every five years.

Consider Information as an Asset

Our final step is to consider information as an asset. This is sometimes easier said than done. Many companies today, however, find themselves in the information business—like it or not. As Peter Drucker warns:

> In the knowledge society, managers must prepare to abandon everything they know.... Knowledge is *the* primary resource for individuals and for the economy overall. Land, labor, and capi-

> tal—the economist's traditional factors of production—do not disappear, but they become secondary. They can be obtained, and obtained easily, provided there is specialized knowledge....

If Drucker is correct, and we think he is, then businesses cannot be content with information systems that merely spew out data.

Systems instead must become flexible, actionable, and able to provide clues to what a knowledgeable workforce needs. Information and people combined become assets.

More Changes to Come

Whether you subscribe to the words of Drucker or you are caught up in the swirling changes in business, technology, and people, it is hard to come away from this topic without thinking that there must be more changes to come. And you are right.

The changes we've talked about so far are also spawning other changes—like how careers are made and how organizational structures are built. In the next chapter, we begin to weave in the changes in work structure and the ways individuals react to change. We are beginning to see organizations organize and reorganize in very different ways. This makes managing business, people, and information that much more challenging.

Chapter 5

Changes in the Traditional IS Work Structure: What's Happening to Your Career?

Over the past few decades, the nature of work, especially white-collar work, has changed dramatically. Organizations in both the public and private sectors have begun enormous gyrations, making organizational changes in efforts to become more efficient and productive.

The downscaling and streamlining of many large corporations have left holes in old organizational structures. These holes sometimes occur in places where knowledge workers once existed. The challenge for business people is to fill the knowledge void without adding people. This challenge becomes even more difficult as many corporations are finding themselves in the knowledge business. That is, to paraphrase management guru Peter Drucker, companies can buy land, labor, and capital, but they have to be "smart" enough to do so.

In this chapter, we discuss the changes in the work structure and how they are affecting virtually all people in all organizations. We'll start by talking about the differences between function-based organizations, process-based organizations, and organizations based on some newer concepts, such as team-based management and the virtual corporation.

We'll also talk about how these new management and organizational scenarios affect information needs. It turns out that, as organizations remold, reshape, and reinvent themselves in different structural forms, their information needs also go through a big change.

We'll close by discussing the changes you can expect to see in a career. What used to be a predictable stair step now resembles hopping from place to place around a ball. You don't always know the next step in your career, and it isn't always a higher one. Career progression just isn't what it used to be.

Redrawing the Lines

People are always trying to do things better. American business people are sometimes preoccupied with the latest management fad to extract a measure of productivity, reduce time to market, or increase the efficiency of a task. For many groups, the search for salvation (or at least, better business practices) leads them to redraw the lines of departments, workgroups, and divisions.

Among the more significant changes is the movement from function-based to process-based organizations. This movement, often spawned by reengineering projects, quality improvement initiatives, downsizing, or other restructuring, moves people out of classic functional roles, such as accounting or marketing, and places them on a mission-oriented team. We'll examine the differences between function-based and process-based organizations in the next section.

Function-Based Organizations

It used to be easy when people joined a large corporation or organization. They knew their place in this world. People would take up a predictable career path. There were slots available in marketing departments, sales departments, finance departments, human resources departments, manufacturing groups, and so on. Everyone knew his place and where it fit in the organization. As shown in Figure 2, the function-based organizational structure aligned people within their various professional functions, such as finance, sales, or human resources.

While function-based organizations have been with us forever, management gurus are beginning to question their value in a make-it-happen-now world. The primary problem with function-based organizations reveals itself when one tries to move anything through quickly. Function-

FIGURE 2 Structure of function-based organizations

based organizations tend to handle projects sequentially—passing work, ideas, or concepts in a predictable order.

Anything that crosses functions, such as the development of a new product, is likely to have to spend long periods of time going up and down a functional area and then finally skipping sideways to the next functional area. Processes such as bringing a new product to market, making major changes, or buying a new business touch many different functions and therefore take a very, very long time to get through the organization. They require special handling at each different functional stopping point.

Jobs in function-based organizations tend to be based on hierarchy. There are workers, supervisors, managers, directors, and vice presidents. Sometimes, in large organizations, there can be even more layers of management. Each job controls a task, but not necessarily a process. Therefore, large numbers of people are required to do small units of work.

A good example of a task-oriented job in a function-based structure is that of an accounting clerk in an accounts receivable function. In traditional organizations, accounting clerks would open invoices, sort them, manually check for proper approvals, signatures, or purchase orders, circle pertinent facts (e.g., date, amount, or terms), and then pass them along for computer processing. The accounting clerk would run a batch total on a group of invoices, band them together, and then send them to a data entry clerk for input into the computer. The data entry clerk would make the required computer entry, run a control report, and compare it to the accounting clerk's batch report. Assuming all was in balance, the data entry clerk would return the invoices to the accounting clerk (or another accounting clerk) for further processing.

The jobs, either accounting clerk or data entry clerk, were task-centered and existed as an assembly-line-type routine. The time for administration, supervision, checking, and control reporting of such a sequential task arrangement added a lot of overhead.

Function-based organizations, however, are predictable, and so are their jobs. Generally, people know where they fit, what their next job is likely to be, and when they may see the next promotion. Their jobs are easy to master and well documented, well known, and well understood.

The problem with function-organizations is speed and efficiency. They can become slow and bureaucratic, developing choking administrative processes. Recognizing this problem, companies began several years ago to experiment with what are called process-based organizational structures in an effort to speed up the lengthy process by which function-based organizations conduct business. We will discuss process-based organizations later in this chapter.

Technology for the Function-Based Organization

Technology needs for a function-based organization are perhaps the easiest to define, because there is a clear orientation towards data. It is easier to get reporting needs and data needs out clearly when all functions are narrowly defined.

For example, different function-based organizations take different views of data. In a marketing organization, most of the activity is centered around a product view of the organization's sales. Classic marketing groups organize themselves around products or brands. They assign a work force to a product, say, blue widgets. That force becomes experts in widgets, especially blue ones. They track widget sales, the sales of competitive widgets, the consumer preferences in widgets, and so on.

By contrast, the sales department of an organization looks at data differently. In the sales view, there's more emphasis on multiple products over a geographical area, whether it be a district, a region, or a nation. The sales group is generally focused on what's happening within their local area or within certain customers' domains. It sometimes doesn't matter to the sales person whether he sells blue widgets, pink ones, or purple. He is more focused on making sure a sale happens.

Therefore, information needs are well defined in function-based organizations. There can be a single emphasis, or view, placed against the data, and most of the data are extracted from databases rather easily. This ceases to be true as we move into the newer organizational structures, such as process-based organizations or virtual corporations.

Process-Based Organizations

In a process-based organization, we cut through the lines of the traditional function. That is, we take representatives from each functional discipline and put them on a team. This is seen frequently in business today as a special project team, sometimes called a cross-functional team. As shown in Figure 3, we just draw the lines of the project team differently.

FIGURE 3 New structures possible in process-based organizations

A process-based team embodies what we sometimes call a Noah's Ark approach. Management will draft two from each species: two from finance, two from marketing, two from sales, two from human resources, and so on. Put these people on a boat and let them set sail to the new world, or, shall we say, the new project goals.

Such a team includes all the skills necessary to bring a new product or concept forward without the need to go through the bureaucratic rules of each individual functional grouping.

Many companies who have tried process-based organization have done so on a special project, for example, to get a new product to market in three or six months rather than perhaps two years. Process teams have proved to be very successful in organizations that have tried this. They are, in fact, quick to cut through bureaucracies, and they can bring decisions to the table faster.

The problem, however, with process teams is that very few of us really know how to manage in a process environment. Who becomes the supervisor, who becomes the team leader, who becomes the person that fills out everyone's performance appraisal? Do you continue to use the process leader or the function leader, who now no longer has direct supervisory capability over the process team, or do you somehow appoint a process leader who really might not know all the nuances of a particular discipline, such as accounting? A marketing person may be inappropriate to judge the relative skill and merit of an accountant's contribution, while the senior accountant may not be close enough to the person's work to adequately appraise his contribution.

These issues are being worked out in companies today. What is important for our purposes, however, is to understand that these changes in organizational structures can and do exist, and, we believe, there will be more rather than fewer of these different structures as the future unfolds.

Implications for Information Technology of a Process-Based Organization

Process-based organizations create a bit more of a challenge for designing effective information technology solutions. In a process-based organization, because you are cutting through all previous functions, information systems must be broader and, sometimes, more complex.

Information systems for a process team, for example, are likely to cut through an entire process, such as the order entry process. In such an arrangement, the information system must be available for the entire order scenario, including, for instance, taking the order; confirming the order;

updating the on-hand inventories; logging the order to the customer's database; fulfilling shipment tickets; and taking the order through shipment, delivery, and billing. So, we can follow through the entire process with an information system. As a result, the system is likely to be more complex and more far reaching.

Another characteristic of information systems needed by process-based organizations is the need for speed. Process teams usually don't have the luxury of time to make decisions and can't wait for information to churn out of a computer. They need fast access to data throughout the process.

Virtual Teams and Corporations

Another new concept for organizations is to draft members with talent that the company does not necessarily possess internally. Sometimes the resulting groupings are called virtual teams or, in cases where the entire corporation is composed of outsiders, virtual corporations. Such teams draw members who are not necessarily company employees and do not necessarily work in the same geographic location.

Frequently, the work of a virtual team or virtual corporation is project based. The team exists only as long as the project does, and then it disbands, members being sent off to join other teams. This project- or mission-focused group is often intensely focused and has very demanding needs for information, coordination, and communication.

Information Technology for the Virtual Corporation

Virtual corporations, by their very nature, add another layer of complexity to the information puzzle. Because not all members of the virtual corporation are company employees, security requirements become a special concern. What information do the non-company employees take with them into a virtual corporation? Should historical sales and shipment data be included? What about competitive activity and intelligence? And what information should be excluded?

Also, in a virtual corporation, there is frequently a requirement for access to far-reaching information sources. Because a virtual corporation is made up of individuals from across disciplines and of non-company employees, it is likely that each member of a virtual corporation will bring in his favorite data sources and private data references. How many accounts are needed for the Dow Jones Information Service, or Dialog, or CompuServe? What are the primary information sources needed by the team? Combining all this information to create knowledge can be a significant challenge compared to either a cross-functional or process-based organization, let alone compared to a function-based organization.

In such arrangements, knowledge-building software, products that border on groupware, or document management software, such as Notes from Lotus Development Corporation, or Folio from the Folio Corporation, can be much more helpful than software packages that rely simply on indexing or database technology.

Geographically Dispersed Teams

With the rise of globalization comes the rise of global teams. More and more, companies and organizations are finding that talent neither carries a passport nor speaks a single language.

Global teams are formed to bring as much talent to bear on a business situation as possible. The good news about global teams is that a large engineering firm making a new pipe fitting can easily have draftsmen in Holland and the manufacturing facility in Hong Kong. The bad news is that global teams are likely to stress the communications capability of an organization to its limits. Transporting information becomes as important as transporting products.

Information Technology Needs of a Geographically Dispersed Team

Geographically dispersed teams require a special kind of communications: communications that allow very free-flowing, unstructured information. They may need to transmit ideas, concepts, drawings, or data. It is critically important that these teams be given several types of information and communication technology as well as the training to use them appropriately. The basic requirements of global teams usually include: fax, e-mail, spreadsheet software, word processing software, graphics software, project-specific data, and database software.

For example, it is appropriate to fax certain types of engineering drawings to the members of a global team for viewing. It is not good use of technology, however, to fax a spreadsheet full of numbers that someone on the other end is likely to have to rekey. A better approach would be to send the spreadsheet as an attachment to an electronic mail document. That way, the person receiving the information can work directly with the numbers and avoid the lengthy process of rekeying the data.

What's Happening to MY Career?

A question frequently asked by corporate citizens today is simply: "What's going on here? What's happening to MY career?" Frankly, that question is more easily asked than answered.

Careers used to be predictable. Figure 4 shows the career progression that one might have expected upon entering a corporate information services group. Job progression used to look like a stair step.

If a person graduated from a credible university or technical school with a concentration in computer science, he could plot out the next 20 years with some degree of precision. He would start out as a programmer, and, assuming the program code he produced was relatively clean (and didn't have too many bugs), he could anticipate a promotion to analyst

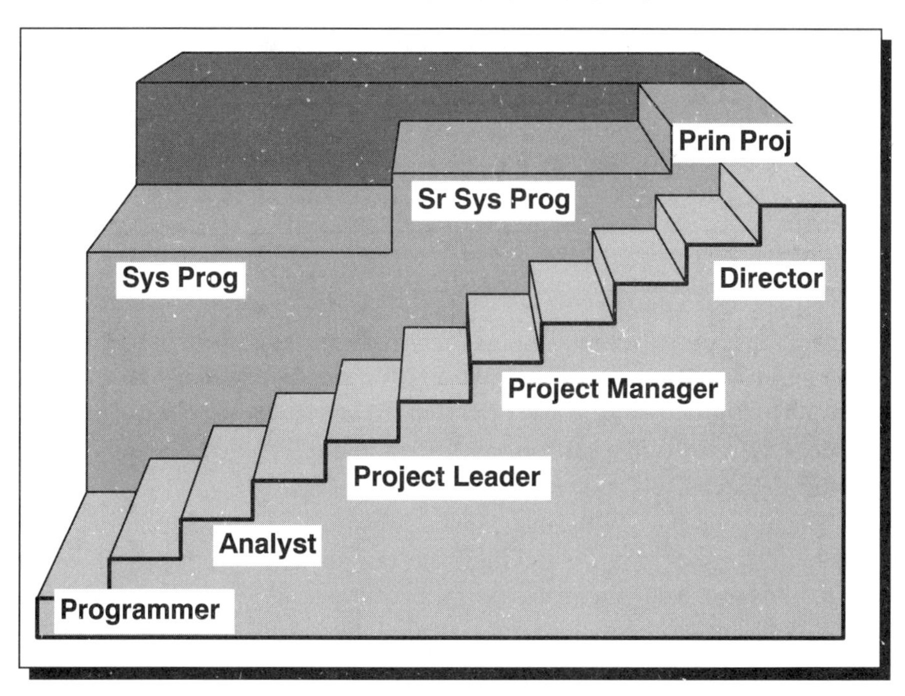

FIGURE 4 Traditional career path

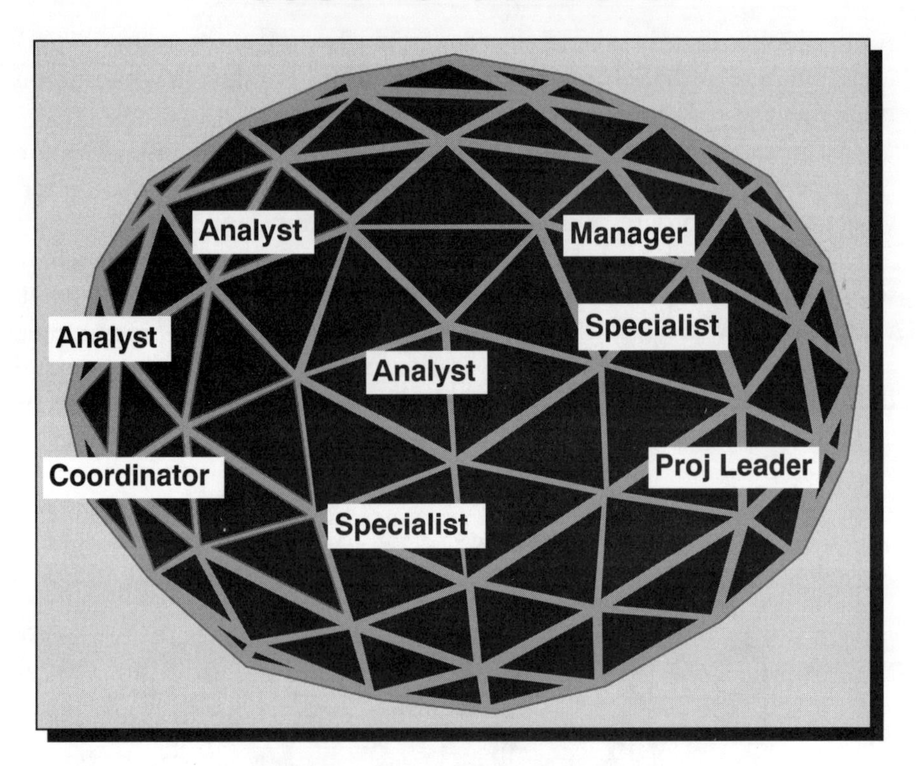

FIGURE 5 Today's chaotic career "path"

within a few years. Then, as an analyst, he would meet a new set of responsibilities. If he met them with some degree of acceptable performance, he would ascend to the next step up the stairs.

Jobs were well thought out and predictable. People knew where they were on the steps and where they were going.

Starting in the 1980s and continuing into the 1990s, the nature of work has changed. Jobs are moving from task orientation to process orientation. This move changes roles, goals, and responsibilities. Now, as shown in Figure 5, jobs come and go in no predictable order. An information services professional (or any other worker) can assume a number of functions without regard to relative rank or title. People may be assigned to teams, stripped of functional or traditional reporting responsibilities, and cast into very unfamiliar work surroundings.

This new, chaotic career path is of great concern to some people. Those who look for a more orderly, stairstep career are often finding themselves out of luck. As companies downsize, rightsize, and realign, jobs take on a very different progression.

Surviving the 1990s-Style Career Changes

The best advice to those looking for guidance in surviving 1990s career changes is to roll with them. It is unlikely that old-style structures will re-emerge in this decade. As people take on more and different responsibilities, new skills, such as knowledge skills, will become important. Out-of-the-box thinking, innovation, and creativity will allow people to participate and find their career strengths.

A final note is that, more and more, the individual must take total responsibility for his own career destiny. As sage management guru, Peter Drucker, recently said in an interview in the Harvard Business Review, "More than anything, the individual has to take responsibility for himself or herself, rather than depend on the company." He added that people must ask themselves, "What do I need to learn so I can decide where to go next?" That, of course, applies to an individual job, but also to an entire career. People, indeed, are responsible for their own information and career destiny.

Chapter 6

The Changing Nature of Relationships with Customers, Suppliers, and Competitors

The business pressures that we cited in Chapter 1—globalization, time to market, and so on—have had a profound effect on how companies interact with their customers and suppliers and on how they compete against their competitors today. Just two or three decades ago, many business transactions took place during face-to-face meetings. Deals were closed with a handshake. In today's fast-paced, intense business environment, the handshake is more likely to take place between two computers.

The advancement of computer technology over the years has brought us a "chicken and the egg" scenario where business relationships are concerned. On the one hand, we deal differently with our customers and competitors because we *can* use technology to conduct our business. On the other hand, we *have* to use technology because our relationships have changed. Often, computer communications are the most effective way to work with clients, supply vendors, and even competitors.

Like it or not, we all have to adapt to living in the Information Age. All too often, this means our society has lost "the personal touch."

Whereas we used to pick up the phone and talk to real live people, we now often encounter voice mail or routing messages. Our financial transactions are frequently completed through an automated teller machine, rather than at the teller's walk-up window. Even your cable television company can send you service messages through your converter box, rather than contacting you via the phone or mail.

In some cases, the Information Age has added a personal touch to business dealings. Companies can now gather and use much more information about their customers. For example, Domino's Pizza keeps a database of "pizza preferences" at each local store. When a customer calls Domino's to order a pizza, the person taking the order asks for the customer's phone number. Based on that information, the person taking the order can retrieve the customer's name, his address, and even what he likes on his pizza and how he wants it prepared. The transaction goes something like this:

Domino's: "Thank you for calling Domino's. May I take your order?"

Customer: "I'd like to order a pizza."

Domino's: "What is your phone number, please?"

Customer: "555-1234."

Domino's: "Yes, Mr. Smith. Do you still prefer a large pizza with thick crust and extra sauce, pepperoni, and green peppers?"

Customer: "Yes, that's just what I like."

Domino's: "And is your address still 123 Main Street?"

Customer: "Yes, it is."

Domino's: "Fine. We'll have that pizza to you in thirty minutes or less, Mr. Smith. Thank you for calling Domino's."

Yes, we all interact differently than in the past, now that computer technology is so pervasive. In this chapter, we'll take a look at the changing nature of business relationships. We'll talk about business trends that companies are following in order to survive and thrive. And, we'll discuss some of the computer technologies that enable (or force) these changes, as well as how your organization can benefit from implementing some of these technologies.

The Changing Nature of Relationships

We have already discussed some of the forces that are driving business changes today: globalization, employee empowerment, variable workgroups, virtual corporations, and more. All of these forces have resulted in new ways to relate to our customers, suppliers, and competitors. Many corporate decisions, particularly those having to do with technology implementation, now include the input of these "outsiders."

Computer technology is helping to create the extended family of the corporate world. A good example is the nuts-to-soup view that Levi Strauss & Company has concerning its business. Levi looks beyond its own walls and actively solicits participation by its suppliers and retailers. The following excerpt from an interview with Levi Strauss & Company Chief Information Officer Bill Eaton reflects that philosophy.

> In the past if someone wanted to create an integrated company from the front to back, you would do it through investment. You'd own the supplier, manufacturer and the retailer. Today you don't have to make the investment. One way to create the vertical company is by exchanging information perfectly, so that the textile supplier knows exactly what we are producing and selling and, therefore, what we are going to need. This encourages just-in-time approaches that make sure they deliver the right amount at the right time. We are doing the same thing throughout the industry supply chain.

As we will see later, Levi is using computer technology to create and manage that vertical company.

Trends in Business That Affect Relationships

There are a number of trends that confuse traditional business relationships. For instance, we can ask the riddle, "When is a competitor not a competitor?" The answer, of course, is "When he is a partner." That goes for customers and suppliers as well. More and more, these outsiders are being brought into the fold of what a company is doing.

Maintaining a business relationship is tricky these days. How do you give the new "partner" enough information to contribute to your business without giving him too much? And, how can you get to know more about your customers and competitors? After all, knowledge is power. The solution can often be found through computer technology.

We are seeing other trends that involve *who* deals with the customer, as well as *how.* As we have already mentioned, the muckety-mucks in the Ivory Tower aren't the only ones making decisions and touching the customer anymore.

Let's take a closer look at what's happening with business relationships today.

Alliances and Joint Ventures

A major trend that we have seen in virtually all industries is the alliance of two or more companies in order to compete more strongly on the open market. Alliances arise from the recognition that no one company can do it all. In many cases, two companies that have complementary skills or product offerings find synergy through cooperation. In a growing number of instances, the alliance partners are competitors that team up to edge out *other* competitors.

Business alliances make a lot of economic sense. The partners have the opportunity to share development expenses, existing or new production facilities, distribution networks, and other scarce resources.

Combining the Best of Breed

One popular form of business alliance is the joint venture shared by two or more companies, where each brings separate pieces of the puzzle to the table. None of the companies can compete in a particular market on the merit of its own products or services. By joining forces, however, each of the companies gains a shared piece of the market.

One prime example of this type of alliance is the recent agreement between Borland International and WordPerfect Corporation. Borland is well known for its database and spreadsheet software for IBM-compatible PCs. WordPerfect has its own stellar word-processing product. Taken separately, these products do not compete against each other.

Borland and WordPerfect agreed to team up to combine these products—the database, the spreadsheet, and the word processor—into a single package called a suite. In this way, the Borland/WordPerfect team could compete head to head against major competitors Microsoft Corporation and Lotus Development Corporation, which both offer multiproduct suites. Without the alliance, both Borland and WordPerfect would probably continue to lose sales opportunities to Microsoft and Lotus.

While the results of this alliance are yet to be seen, there clearly has been an impact on the way Borland and WordPerfect relate to each other.

The partnership is encouraging (forcing?) them to share development strategies and techniques. Both companies are committing to a future that will mesh more than product.

If You Can't Beat 'Em, Join 'Em

Tough economic times and a global economy are two leading factors in the drive for competing companies to unite for the common good. Ray Noorda, chairman of Novell Corporation, calls this "coopetition." Perhaps no industry illustrates this trend better than the automotive trade.

With American auto companies finding it difficult to compete in foreign (and even domestic) markets, the Big Three have turned to joint ventures with their competitors. Chrysler Corporation has teamed with Mitsubishi, General Motors is allied with Isuzu, and Ford Motor Company produces products in concert with Mazda. Can alliances among Ford, Chrysler, and GM be far off?

The trend isn't limited to American companies, either. Hyundai of Korea and Mitsubishi of Japan have paired up to stave off the competition. (This venture has more than the corporate competitive spirit to surmount. The two countries' politics are a real burden to overcome in this case.)

These examples, and countless others like them, show the measures that companies are willing to take in order to gain or hold onto market share. It is a bold step to team up with "the enemy," but sometimes it is the only way to survive.

Getting to Know You

Companies today have the opportunity to know more about their customers than at any other time in history. We have elevated the process of collecting and analyzing data about our customers almost to an art form. And all of this information gives rise to a more intimate type of relationship. To wit, even the pizza maker knows that you like mushrooms on your Friday night pizza, even before you place your order!

In 1992, we saw an early attempt by the Kroger grocery food chain to gain more insight into its customers. This particular program shows that companies can easily learn more about their clients by employing technology.

In this example, Kroger issued "gold cards" to regular customers. The card was merely a means of identifying a particular shopper: name, address, and various other demographics. When the shopper went through the checkout process, the card was scanned for the customer information. Next, all of the groceries were scanned, as usual. As gold card holders bought specially designated products at the Kroger stores, they collected points that could be exchanged for gifts and merchandise. More important to Kroger, the store now had very detailed information about the customers and their shopping habits. The store manager now knew what brand of toilet paper John Jones was likely to buy, and how much dog food Jane Doe's household uses in a week. That's powerful information to a retailer. This kind of personal information can really shake up the traditional grocer/shopper relationship.

Tearing Down the Outside Walls

Another business trend that we are hearing a lot about these days is reengineering. Many organizations are looking at their internal processes and streamlining the way things are done. Companies that have been most successful at redesigning their work flows have not stopped with internal processes—they also look at the flow of information outside the company. Key customers and suppliers are being asked to contribute ideas for and participate in the operation of new processes.

Ford Motor Company reengineered its accounts payable process years ago, with dramatic results. Under the old system, Ford employed more than 500 people who pushed paper to get the bills paid. The process went something like this:

A Ford purchasing agent issued a purchase order to buy parts or supplies. A copy of thc PO went to the accounts payable department. When the goods arrived, the material control department sent a copy of the receiving document to accounts payable. At the same time, the vendor was sending an invoice to accounts payable. It was the job of the accounts payable personnel to match the PO, the receiving document, and the invoice. If all three matched, the bill got paid. Frequently, however, there were discrepancies, and accounts payable had to track down the differences, resulting in delayed payments and higher costs.

Ford undertook a radical reengineering of this process. An online database replaced most of the arcane paperwork. The company included its suppliers in the new process, enabling the vendors to eliminate printed invoices. All purchase requests and material receipts were stored online, and a check for the vendor was cut automatically as goods were received. As a final result, Ford was able to reduce the size of the accounts payable department by more than 75 percent.

A significant factor in the success of Ford's new process was the participation of the vendors. The relationship between Ford and its vendors became less adversarial and more cooperative, all in the name of efficiency.

Retailer Nordstrom, Inc. is another firm that strongly believes in partnering with suppliers. In fact, data processing manager Charles Mitchell calls it the key to success in retailing. The company's VIP Express e-mail system is used to give suppliers fast and easy access to Nordstrom's buyers. Mitchell feels that this system has helped Nordstrom strengthen its relationships with suppliers.

The moral of these stories is that companies should look beyond their own walls when looking for better ways to do business.

Power to the People

Empowerment: It's the new buzzword in business today. It refers to management's passing the decision-making process on down to the worker bee level. The goal is to have decisions made at the very first level of initial customer contact.

Perhaps nothing has had an impact on business relationships as much as empowerment has. Since the customer works directly with the front-line employees, there is no need to consult management over small issues or decisions. The result (usually) is a faster transaction and a happier customer.

Our old friend Levi Strauss & Company is noted for its philosophy on empowerment. What's more, the company's efforts extend beyond its own employees to its suppliers and retailer customers as well. The company—along with its whole supply and sales chain—uses information technology to allow local decision makers to make the choices that affect market research, manufacturing, marketing, distribution, and sales. We'll take a look at how Levi Strauss accomplishes this feat later in this chapter.

Workgroups

As we mentioned in Chapter 5, there is a trend away from traditional hierarchical reporting structures and toward cross-sectional workgroups. Companies form teams of people that evolve by job function, not by departmental or geographic lines. They represent the ultimate democratization of business; people have equal access to information and to each other.

Workgroups tend to form and realign on a project-by-project basis. Therefore, workgroups are fluid, acquiring and using people resources as they are needed, and releasing them when they are no longer needed. The shifting nature of the workgroup affects how companies interact with customers, suppliers, and competitors, and especially how employees within companies relate to each other.

Manufacturers, in particular, have found benefits in forming product teams that include people from all areas of a company. This grouping of employees becomes responsible for a product, from start to finish, from design to sales and marketing.

Information Technology and Business Relationships

At the beginning of this chapter we talked about the information technology "chicken and egg" scenario. The point is that technology and computers serve two ends: they *enable* the change in relationships, and they *help us to cope* with the change in relationships.

There is little doubt about the benefits that computers provide when it comes to communicating with our business partners and responding to customer needs and competitive pressures. We have access to more information than ever before, and it can be sliced, diced, chopped, and served in more ways than ever before. What's more, we can carefully structure an outsider's access to data. For example, a manufacturer may give its retailers online access to product pricing without allowing access to cost information. The manufacturer gives a trickle of pertinent information without opening the floodgates.

The intelligent use of technology can help us to serve the customer better and to beat out the competition. Let's take a look at a few of the computing technologies that are available and how they may be implemented to improve our business relationships (and, hopefully, the bottom line).

Electronic Data Interchange—EDI

Not since the invention of the telephone have we seen a technology that has impacted business communications as EDI has. Strictly speaking, EDI is the exchange of information—usually raw data—between computers. Very little human intervention is needed.

With lightning speed, data—product orders, inventory levels, sales figures, etc.—move from one organization to another. Select information is exchanged in a predetermined manner. Note the importance of the phrase "select information." With EDI, you are giving your partner a very specific set of data. What you are *not* giving is uncontrolled access to your information base. That's what makes EDI such a good means for communicating with outsiders.

Many of *Computerworld Magazine's* Premier 100 companies, all top users of information technology, are concentrating on creating new systems or enhancing existing systems for EDI in 1993. The efforts are toward streamlining production and order fulfillment, improving communication with customers, and distributing information more effectively.

Many large companies have implemented EDI within their own domains. Data are transmitted from different divisions, subsidiaries, and widespread locations. But the most expedient use of EDI technology may be outside a company's borders, in communicating with customers, suppliers, and perhaps even competitors. Let's take a look at how Levi Strauss has chosen to use this technology to create an empowered employee base and supply chain.

Levi Strauss is accomplishing information exchange on a grand scale through EDI, extending the process to its retailers and suppliers. The company set up its first EDI system for retailers in 1986, making it low cost and providing the users with adequate training to promote acceptance. And accepted it was. The retailers use the system to track inventories and maintain optimal levels of product stock, among other things.

As for the suppliers, Levi Strauss simply said, "We are your customer, and this is how we want to do business." The company has worked with its suppliers to get most of them on board for the electronic exchange of data.

A key to the success of the program was to help set and enforce communication standards. Levi Strauss became active in the Voluntary Industrial Communication Standards Committee, which set up the ANSI X.12 standard for the consumer products group. Now everyone along the supply chain can adhere to this standard and communicate effectively via computer.

The health care industry is another trade that is moving rapidly to EDI. With the pending reforms in the health care system, EDI holds the potential to help the whole industry save as much as 35 billion to 50 billion dollars annually. Moreover, health care providers will have the necessary information to make better decisions that will lead to better health.

Leading the effort for adoption of EDI in health care is a coalition of government and private organizations known as the Workgroup for Electronic Data Interchange (WEDI). It is pushing for widespread use of ANSI X.12 interface standards to process claims.

Nonstandard forms and proprietary, unintegrated technologies are the barriers to EDI. Once these obstacles are overcome, the rest is easy.

Electronic Mail

Both public and private e-mail systems are being used to expand the realm of communication in business. There is virtually no limit on whom you can talk to. We have referred to e-mail as "the great equalizer" for just that reason. E-mail can overcome the chain of command and has a way of circumventing even the most protective secretary.

Most large organizations have implemented e-mail for use within their own ranks. There are also many options for exchanging notes and other documents with people on the outside. CompuServe and MCI Mail are two of the most popular subscription services for the public exchange of e-mail. Internet is also a widely used service for e-mail. These and other services like them provide for quick and easy global communication.

New on the e-mail scene are wireless systems based on radio communications. Now it is not even necessary to have direct connect or dial-up access to your e-mail service. With radio communications, your e-mail can find you no matter where you are.

You can easily see the impact that easy e-mail can have on business relationships. Informal communication can take place in an instant. What's more, e-mail makes it possible to cut through thick layers of chain-of-command bureaucracy.

As a testament to the power of e-mail, Currid & Company conducts a large portion of its business with this technology, both internally and externally.

Telephone Systems and Intelligent Networks

Telephone systems and intelligent networks represent the marriage of two technologies: evolving communication tools and distributed computing systems. With both, we have come to look on our telephone systems as more than just a way for two people to hold a conversation.

For a simple implementation of the technology, consider the way that telephone systems can be used to screen calls and route callers to an appropriate destination. For example, how many times have you encountered a routing service that tells you to "press 1 for so-and-so, press 2 for such-and-such," and so on?

This is certainly an "interesting" approach to customer communications. We say interesting because the same approach can be either a help or a hindrance to your business. Customers sometimes like the efficiency of accessing information through the phone. Consider, for instance, checking on your bank account balance. At other times, customers hate the experience of listening to so many messages and pressing so many buttons. One of the authors recently tried to learn the times and cost of a movie showing at a local museum. It literally took ten minutes to drill down to the desired information. How frustrating!

Your telephone (along with the unseen network of computers employed by your service provider) brings you many other information-processing services. Consider 800 services, call forwarding, personal numbers, voice mail processing, speech recognition, fax messaging, remote network management, and cellular telephones. All of these services give you a myriad of ways to communicate with—and to relate to—your customers, suppliers, and competitors.

Mobile Computing

Computing technology that can be taken on the road opens up whole new business opportunities. From the traveling salesman to the bedside physician, people can use portable computers to access data from previously inaccessible locations. The key to empowerment is the ability to assemble the information that is needed for a decision or action on the spot, and mobile computing is a means to that end.

Sales representatives who travel to their customers' sites have found mobile computing to be a real boon. A salesman can now set up a small computer in his client's office, give his whole presentation with the aid of computer graphics, check product inventory levels by dialing into the company's central computer, and place the client's purchase request on line. After the sale, the salesman can generate a report for his divisional manager while sitting poolside at his hotel and then send it to headquarters via e-mail. Privately, he can tally his expected commission in a spreadsheet. If it's all done right, there's no need for cumbersome presentation materials, excessive paperwork, or long lead times for orders to be placed and processed. And best of all, the customer is happy about the efficiency of the new process.

There are a thousand and one uses for mobile computing, especially in areas where it was not practical to use desktop computers. Imagine your doctor visiting your bedside in the hospital, accessing treatment information via a hand-held computer. Or your personal banker coming to your office during your lunch break to wrap up your new mortgage application. Or the building contractor giving you a detailed shopping list of materials needed for that addition to your house.

In short, mobile computing has the ability to localize decision making, which in turn brings us closer to the customer.

Still More Technology

EDI, e-mail, phone systems, mobile computing: These are just a few of the computing technologies that enable changes in our relationships with our customers, suppliers, and competitors. There are many others, some of which will be discussed in some detail in the latter half of this book. Just imagine the impact of these technologies: facsimile, UPC scanners in retail outlets, groupware, computer-aided manufacturing, multimedia, expert systems, and imaging. The list goes on and on.

As for the benefits to our relationships, they are almost as endless. We can respond faster to the customer. We can give him better information. We can reduce the time to conclude a transaction. We can communicate around the world in seconds. We can eliminate (or at least reduce) time-wasting processes and activities.

In short, our business world is changing. Computer technology both enables and drives the changes in our business relationships. Technology provides us with innovative ways to extend our virtual corporations. In the next chapter, we'll look at what you can do to prepare for the future in this new business world.

Chapter 7

Change and What You Can Do to Prepare for the Future

Computers change everything. If you don't believe it, just stick around a little longer. Once they are properly installed and their technology is correctly implemented, computers can change much of the business culture. Computers can help people communicate more quickly and pass information around an organization efficiently. Technology provides a tool that lets people increase customer service levels, act quicker, and get smarter. The changes, however, don't stop there.

Computers have profound effects on individuals, on their jobs and responsibilities. We believe that people, whether they consider themselves computer people or business people, need to prepare for the inevitable computerization of their careers. The impact can be great, and more importantly, coping with the changes may be more difficult than many people think. The biggest mistake people make is thinking that change will be easy and that they don't need to prepare much more than desk space for a new computer system.

In this chapter, we talk about change. Many people don't realize how impactful computer changes are. Effective systems uproot the status quo, changing corporate cultures as much as they change business processes.

Ineffective systems can muddy up company processes and procedures and bring them grinding to a halt.

Some people are woefully unprepared for what happens—even those people who consider themselves computer professionals. We'll identify two psychological models for how people deal with change and then map them to the computer change experience.

Then, we'll get job specific. We'll identify a number of white-collar professions and begin to map out advice and tips for getting computer literate.

It is important for people—including you—to recognize that computers really do change everything. Some things will get decidedly better, other things will not. For sure, computer systems can cure many corporate ailments, but they have yet to cure the common cold.

To be successful, one must recognize that changes will occur. Oftentimes, the changes will require addressing the human side of an organization—redefining roles and responsibilities. Simply plunking a couple of PCs down on people's desks is not the beginning or the end of a company's computerization. It turns out that not everyone will deal positively with computer-induced changes, especially if some feel their careers and jobs are threatened.

Change Is a Six-Letter Word

As children, many of us were cautioned against developing a vocabulary of four-letter words. It just so happens that the English language is sprinkled with a rich set of single-syllable words that happen to have four letters—and Mom didn't like us using them. We are taught to avoid these words, not to say or think them, and many of us slip up and say them only occasionally.

The word "change" isn't one of them. It is a single-syllable, six-letter word. Mom didn't forbid us from saying it, although some people act as if she did. In business today, the word "change" falls right into the category of the feared and dreaded. It is a word and a process that people want to avoid. But, some changes can't be dodged forever, and, we think, computers will inevitably move onto the desktops of most business people.

As computers are accepted into organizations, their impact is not always seen overnight. Aside from applications that focus on a specific business task, such as taking orders or paying bills, organizations don't always see the impact of computers right away. Over time, changes do occur. Jobs are created, jobs go away, roles and responsibilities change.

Sometimes these changes are predictable, sometimes not. For example, a computer application such as e-mail may look on the surface like a moderately impactful application. E-mail allows people to leave each other messages so they can avoid telephone tag. But, telephone tag avoidance is hardly the tip of the iceberg. It turns out that successfully implemented e-mail can actually help flatten out an organizational hierarchy. Since e-mail messages travel through unfiltered wires, e-mail lets everyone communicate with everyone else without regard to rank or title. Some people call it "secretarial bypass" because it makes executives and decision makers so accessible. It also lets people violate the laws of time and space and keep a dialog going no matter where on the planet people happen to be.

Other applications, such as decision-support systems or executive information systems, can put important information into the hands and heads of key decision makers quickly, often without somebody filtering out the good news or bad news. People don't get in the way of information. The changes ushered in by new information systems just aren't predictable, especially when people focus primarily on the specifics of a computer application.

We come back to computer-induced change. It isn't predictable. Moreover, no matter what change takes place, the process of change is often uncomfortable. Resistance is natural, even if the change is thought to be a good one. Change can be messy and even cause people to become dysfunctional.

Academicians and organizational psychologists have been studying reactions to change for many years. In her book, *In the Age of the Smart Machine*, Harvard Business School Associate Professor Shoshana Zuboff uncovers resistance to change in organizations. She cites numerous cases where workers and supervisors alike grappled with the changes brought in with computers. Feelings emerge from almost every direction, from "we don't trust the computer" to "I am losing my power."

Among the best work that has been applied to the computer-induced change phenomenon comes from Daryl Conner, of the Atlanta-based company Organizational Development Resources (ODR). Conner applies two models to how people react to change. The first, when the change is perceived as positive, shows a five-stage response. The second, when change is perceived as negative, elicits an eight-stage reaction.

Coping with the Good News—Positive Reactions to Change

Even when change is thought to be good, people can become dysfunctional for periods of time. This usually happens when expectation levels rise far beyond reality, and nothing, even the magic of a well-implemented computer system, can meet inflated expectations. Then when reality sets in, hope is dashed and people become unhappy with the computer system, even though it may be better than the processes it replaced.

Conner's five stages of positive response to change, as shown in Figure 6, outline the moods or phases that people encounter:

1. Uninformed optimism
2. Informed pessimism
3. Hopeful realism
4. Informed optimism
5. Completion

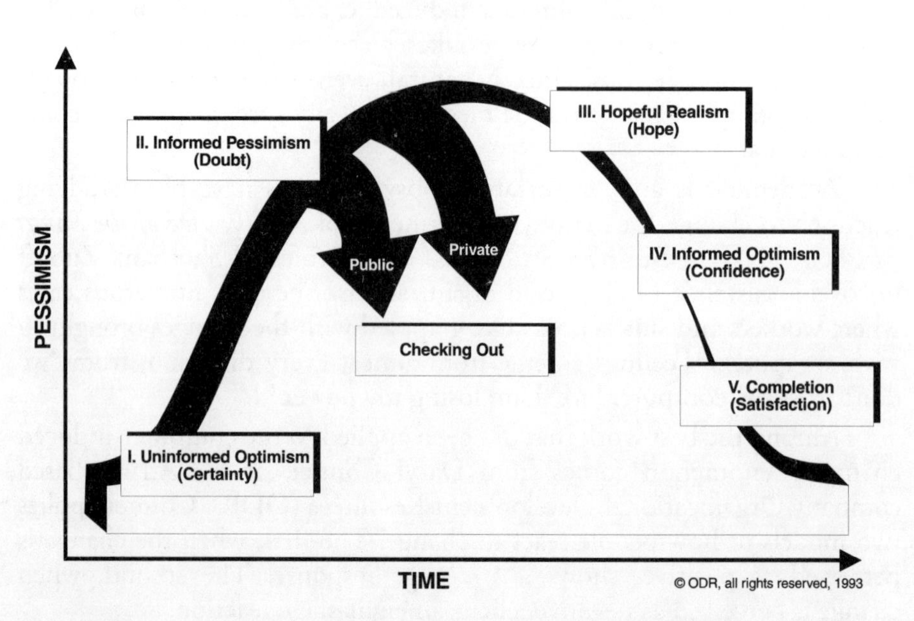

FIGURE 6 Conner's stages of positive response to change

Consider the mythical company, Worldwide Widgets, Inc. as they deploy a new inventory tracking system. Here's how people can experience the change process.

Stage 1—Uninformed Optimism

Managers and staff are excited about the installation of a new computer system. They have great expectations about how business will run better when it is installed. For years, actual inventory and the computer never agreed. The new system promises to get everything in sync, produce perfect pick tickets for the warehouse clerks, and deliver dandy reports for management. There's going to be something for everyone, and life will be wonderful as soon as the new system is installed. On day one, everything goes smoothly—orders are entered, shipments are shipped.

Stage 2—Informed Pessimism

The system has been installed for a week or so, and the honeymoon is over. Coding snafus mistakenly counted purple Widgets as pink ones, and an order to a major customer was sent out wrong. The warehouse clerks stop trusting the system and threaten to go back to manual paper pick tickets. What's more, those promised dandy management reports are dreadful. They don't divide widget shipments across correct geographical alignments, and so the sales manager in the New York region can't figure out where her customer orders are going.

At this point, an insidious problem can surface. It's called *checking out*. People, either publicly or privately, give up on the new system and adopt some other course of action.

In the case of our widget inventory system, the warehouse clerks who threaten to return to their manual paper-intensive process are publicly checking out. They are rejecting the new system and making a big issue of it.

On the other hand, if the New York region sales manager stops reading her reports and quietly instructs sales representatives to audit customer shipments manually, that would be another form of checking out. In this case, she does not surface the problem, but she is taking action that undermines the intent of the new system.

Public checking out, which involves people expressing their problems, can be dealt with. When the warehouse clerks complain, something can be done. Either the programs or the processes can be modified.

Private checking out, where people don't complain but begin to take alternate measures, can cause problems. People can't fix a problem they don't know about. If the New York region sales manager doesn't complain, no one will know that the new system's reports aren't fulfilling her needs.

Stage 3—Hopeful Realism

The third stage comes as pessimism subsides. Assuming that problems identified in the second stage are surfaced and resolved, people begin to settle into using the new system. To arrive at this point sometimes requires corrective actions, such as fixing programming bugs, creating new management reports, or retraining people on the system.

As people work more with the system, learning of its strengths and weaknesses, they gain a clearer picture of its value to the organization. Warehouse clerks figure out how to balance inventories properly, and sales managers find a new report for widget sales by region. More often than not, its value turns out to be less than imagined in stage one but better than feared in stage two.

Stage 4—Informed Optimism

At some point in the process, people begin to take ownership of their new system. Once they've worked through the bugs, wrung out reports, and moved their day-to-day business activity into the system, they begin to find new uses for it.

Perhaps that sales manager can now figure out a way to take the customer shipment information and use it to sell new widgets. She might check a customer's purchases from one promotional period to the next or compare a competitor's activity. She begins to use the information supplied by the system in ways that she may not have thought about previously.

Stage 5—Completion

The final stage, completion, is characterized by a return to normalcy. People have accepted and adopted the new system and assimilated their business activity to it. At this point, people may well have forgotten their old processes and cannot imagine what their work was like before the new system.

A good example of technological change that has been completely adopted in organizations is word-processing technology. Few people remember what it was like to edit a report or correct a spelling error before modern-day word-processing software. The process was pretty painful. Secretaries kept desk drawers full of fix-it material—usually pasty white paint for small corrections and scissors, white tape, and glue for larger jobs. It was time consuming, messy, and not fun.

Coping with the Bad News—Negative Reactions to Change

So much for the good news. Sometimes computer systems are not perceived as a good thing. New technology can threaten roles and goals, status and status quo. People can and will resist anything they perceive as threatening.

Bad news, or news perceived as negative, is rarely met with cheers and handstands. Instead, people deal with it either publicly or privately by passing through a number of phases. Often these phases are predictable and map well to the eight-stage model identified by ODR's Daryl Conner.

Conner's model is an adaptation of a model originally developed by Elisabeth Kubler-Ross, who was studying terminally ill patients. Conner observed that people in an organizational environment exhibited stages of emotion similar to those seen in clinical environments. Being the target of change, whether from a physical cause or from a job-related one, puts people in the mode of having to deal with it.

Conner's adaptation, as shown in Figure 7, depicts the emotional highs and lows that people experience when encountering a change they perceive as negative. Each of the phases, explained below, can be mapped to activity seen as computers begin to change jobs.

Let's go back to Worldwide Widgets, Inc. and observe people in the accounting department as they learn about a new system for handling accounts payable. This time employees are not meeting the new system with good feelings. In fact, people in the accounting department, from clerks to the supervisor, don't want the new system.

Stage 1—Stability

The first stage is that of stability. The status quo is firmly entrenched, and processes are set in concrete. People know their jobs and proceed with them.

At Worldwide Widgets, invoices are processed in a controlled manner. When an invoice arrives, the clerk checks the bill and simply circles important information. He bundles several invoices into batches, runs an adding machine control tape to total the invoice amounts, and then passes the bundle to a data entry person. The data entry person keys the information into a computer terminal and passes invoices, adding machine tape, and a computer-generated control report back to the accounting clerk.

Everything is proceduralized, documented, and functional until, one day, the manager calls a meeting and announces a change. A new

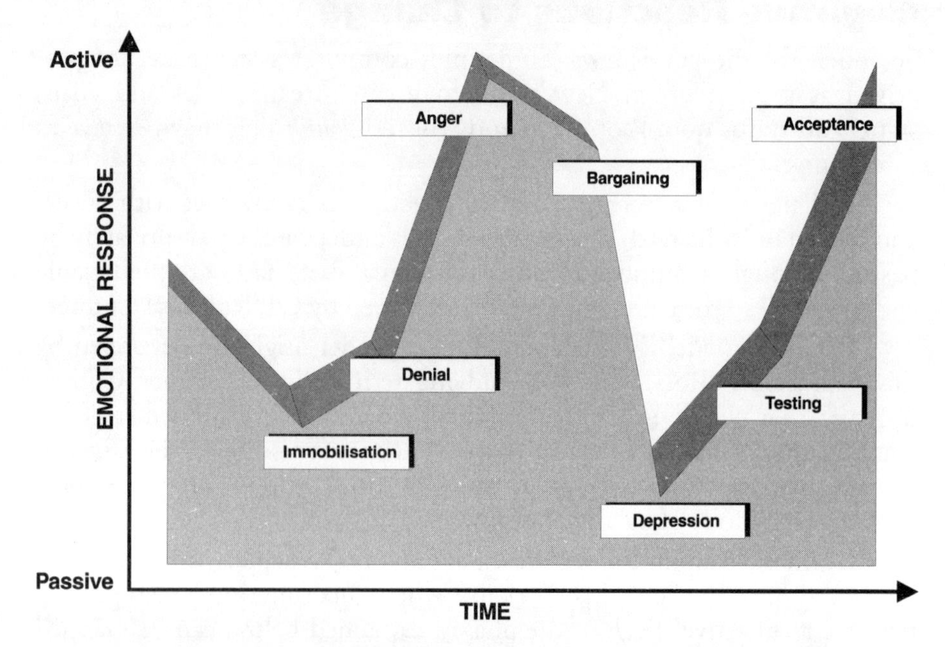

FIGURE 7 Conner's stages of negative response to change

computer system will be purchased to handle invoices more efficiently. The new system will involve a one-step process whereby accounting clerks enter all information into the computer and data entry people are no longer needed.

Stage 2—Immobilization

Immobilization sets in first. Staff members, many of whom have been at the job for years, cannot believe their eyes and ears. Accounting clerks, unfamiliar with computer terminals, fear their new responsibility. After all, they are accounting people, not computer people. Likewise, data entry clerks no longer have a part of the process, and the handwriting is on the wall indicating they won't have jobs much longer.

At this point, people cannot move. They are shell-shocked by the news of the new system, roles, and responsibilities. The change may be so alien that the people involved simply don't understand what is happening.

Stage 3—Denial

Unfreezing from the shock of immobilization, people begin to reject the notion of change. Commonly, they say, "That system may be fine for other people, but it won't work here. We're different." Sometimes they negate their part in the process of change and try to ignore it, hoping that it will go away.

Stage 4—Anger

Another common reaction is to become angry and lash out. Often outbursts of anger are indiscriminate and are directed at anyone who happens to be there. Anger may or may not be seen in the workplace. If it is, it is usually directed toward someone who is trying to be supportive, such as a coworker.

Stage 5—Bargaining

At some point, people try to neutralize the negative impact of the change by negotiating their role in the new environment. Data entry operators may start looking for another job that is largely consumed with inputting information into a computer, while accounting clerks may begin to examine how they could modify the new procedure.

According to Conner, bargaining is the first phase of acceptance of the change. The earlier phases involved different forms of denial, but now people are confronting reality and beginning to try to find their place.

Stage 6—Depression

As people begin to deal with the impending new system, they feel victimized and carry a sense of failure. They may think: "This is just another management ploy to save money, at the expense of our jobs." Or, they may think: "Maybe if we could have processed more invoices per hour, or if we had fewer errors per batch, we wouldn't have to deal with this change."

While the depression stage can be agonizing, it also has a positive effect. People are acknowledging the change.

Stage 7—Testing

At some point, people begin to find their place in the new environment. They explore ways to contribute. Our accounting clerks may choose to take a night course in typing or basic computer skills. The data entry operators may begin to shore up other skills and look for work as administrative assistants.

Stage 8—Acceptance

Finally, people begin to deal realistically with the changes. They may not like the new system, the new procedures, or their new roles, but they accept them. They find ways to become productive in the new environment, even if they have to change jobs.

According to Conner, there is no guarantee that people will get through each stage by themselves. They may need help, support, or counseling. If someone gets stuck in a stage, she could exhibit so much dysfunctional behavior that it will require special help.

Change Happens

Every organization will find its own methods of dealing with the inevitable change process. What's important is to anticipate that reactions and resistance to change can and will happen. As much as possible, people should prepare to help out coworkers during times of change.

How We See Ourselves—Computer People versus Business People

What are you? A computer person, or a business person? Depending on your personal background, that question may not be so easy to answer.

In today's world, we still see a distinction in many organizations between computer people and business people. This is especially truc in large companies.

For example, there are computer people. These are designated people in companies that service computers or the information needs of other people. Most often, we call these information services (IS) people or management information services (MIS) people. These people create, protect, or preserve information.

Then, there are business people. These are generally workers who use information. They are, in fact, responsible for taking that information and turning it into knowledge.

As we see the future unfold, the distinction between computer people and business people will begin to blur. In fact, it already has in some organizations. In small to midsized businesses and in new organizations, frequently there are no information services people or only a handful of people dedicated to the task of keeping computers going. Instead, business people, or what once were computer people, are infused into the organization servicing business functions and business processes.

Likewise, in large organizations, things are beginning to change as companies begin to infuse business teams with information services professionals. This is sometimes called decentralization, a process not always welcomed by the information services professional, who is relocated to take office space directly in a business department. Such professionals may view their future career choices as confused.

Frequently, these moves start with applications development people, such as programmers or analysts. That is, people formerly tasked with developing computer applications for business functions are now assigned to offices within the business unit itself.

When these people are sent to the various functional or process units, they often come under a different reporting structure than they once had. The new reporting structure frequently involves the technical person reporting directly to a functional manager, such as a marketing manager, sales manager, or financial manager, instead of to an information services manager.

We believe that over time, as technology unfolds, there will be more occasions to associate information services people or computer people with business people and imbue them with normal business processes. However, the reporting structures can run either back to the information services group or to the functional or line-management people, depending on the corporate culture. The corporate culture appears to have more to do with the success or failure of these forms of organization than does any other single element.

Whether you consider yourself a computer person or a business person, we can identify the different tactics you should take to prepare yourself and your individual career

Computer People

Computer people, those who have been tasked with developing information for the rest of the company, have a number of challenges ahead as new technologies are introduced. The first of these challenges is to let go of what used to be classic information services tasks. That is, there is no longer a requirement for the information services person to take 100 percent responsibility for all the company's information and interpretation of that information.

Specifically, the computer-focused individual needs to learn new tools for the development and maintenance of company systems. More specifically, these tools revolve around three areas: networking, database management, and applications.

Networking

In networking, it is important that information services people be well schooled in connectivity, especially in local area networks (LANs) or wide area networks (WANs)—that is, systems that deal with passing information packets around an organization.

Schooling in the networking discipline often involves taking courses outside the normal academic patterns. Many colleges and universities have yet to establish a complete enough curriculum to support the education needs of networking professionals. Specific courses should include: basic networking principles, network design, bridging and routing technologies, hardware components, and network operating systems.

If your company uses Novell's operating system, you can choose among several third-party providers of network training. Their course offerings span the gamut from basic networking infrastructure all the way up to managing a network. Courses are available for other operating systems, such as Banyan Vines or Microsoft LAN Manager, but they are not so plentiful as for the Novell environment.

Database Management

As companies deploy new technology, they are likely to bring in new database tools and structures. Simply stated, they use databases as an organized repository for the company's key data elements. In more cases than not, we believe, the database tools will use the client/server model, which calls for the desktop computer and a server to divide the work of an application.

Generally speaking, in large companies, one person will be developing the back-end database, and another, using the front-end tools. In a smaller company, this is not the case. A single individual will, in fact, be generally tasked with aligning the data and then writing computer applications to get the data into and out of the database.

Specifically, for the information services person, training in database concepts, database technology, and back-end and front-end tools is a requirement. Individuals who are tasked with maintaining and preserving the back-end database are well advised to take vendor-specific courses for the particular database product that the company has chosen.

For example, if your company has chosen Oracle Server as its standard database server, then, certainly, courses in Oracle will be required for optimal database design and setup. People who will be working on the database engines themselves will need additional courses to pick up a good grounding in database theory. For example, courses will include: "What is relational database theory?", "How to use SQL or structured query language," and "How to set up and normalize data properly."

Applications Development

For individuals who are assigned applications development activities with database tools, it is important to learn generic database concepts as well as how to use specific tools. In this case, it is important to take courses first in subjects such as SQL and generic database design, and then in a specific tool, whether it be PowerBuilder, Visual Basic, or some other tool. These courses are taught by the vendor, a third party, or sometimes even a local community college or university. In some cases, quick training sessions such as two- or three-day intensive courses can benefit the individual by getting him or her quickly immersed in the use and nuances of the tools.

We believe that, as tools become easier to use and the concepts that underlie application design becomes better known, there will be more courses on the business applications than on the tools. The tools, however, will become easier to pick up without a tremendous amount of training.

Business knowledge is sometimes difficult to acquire. It is not always taught in a two- to three-day course or available in the curriculum of a college or university. Instead, business knowledge is frequently learned by actually working on the job and by reading trade periodicals and business journals.

Business People

For the purposes of business people, a great deal of change will occur as the company embraces computer technology. Business people generally become more self-sufficient, more responsible for their own information destiny, and more responsible for the acquisition and interpretation of the information they have to use in their jobs.

Business people must become more computer literate so that they can begin thinking for themselves. In the next section, we will detail a number of business professions and describe the personal responsibility for information that people within these professions must acquire. In some cases, people will already be gravitating toward this type of knowledge. That, unfortunately, is not universally true.

Business Roles and Information Responsibilities

We believe that business people's information responsibilities change dramatically as computer technology becomes a part of an organization's corporate culture. As mentioned previously, the overall effect is that everyone becomes more computer literate. But, depending on the specific role an individual has in the organization, the relative amounts of mechanical versus concept computer literacy will, in fact, vary.

According to Sproull and Keisler, the differentiation between people who use computers and people who do not use computers will cease to exist within the next five years. Everyone, they believe, will become computer literate.

We agree with the assumptions of Sproull and Keisler, but feel that the timetable may become longer, perhaps 10 years. We believe, however, that computer literacy will spread at an accelerated pace in organizations where the company culture dictates that people assume more information responsibility.

Already, we have observed companies where computer literacy is not an option but is mandatory for employment. It would be difficult to employ a corporate accountant who didn't know how to use an electronic spreadsheet, such as Lotus 1-2-3 or Microsoft Excel. Likewise, who would hire a secretary who didn't know word-processing software? How helpful would this secretary be without knowing how to use a spell checker, or if she had to use whiteout each time an edit was necessary?

Surely, people without certain computer skills are already excluded from many professions. We believe the trend will continue and spread to most white-collar occupations within the next two to five years.

Because computer literacy is becoming such a requirement, we offer the following section to highlight certain positions and the amount of information responsibility (and computer literacy) each position will eventually require.

Executive Managers

Executives have long been held hostage by information-laden subordinates. In many organizations, the executive knew only what she learned through subordinates' research. In some cases, unfortunately many cases, the executive was not always informed of all the facts. Executives were often painted a rosy picture of business prospects, even when things were starting to sour. Consequently, bad business decisions were sometimes made.

Executives in the dynamic organization will become far more computer literate than they have ever been before. They will be capable and responsible for looking up their own information, making the proper analyses and interpretations of that information, and challenging the wisdom of decisions made by colleagues and subordinates.

Frequently, executive computing starts with electronic mail. That is, even though executives are sometimes not very keyboard literate, the ability to send and receive small bits of information very quickly brings executives to the keyboard. Beyond electronic mail, the next most important executive computer activity is that of performing queries on company data. That is, the executive, when armed with some computer literacy and position-support or query tools, should, in fact, be able to pull information along a flexible route. This may be sales, orders, inventory, or other company-specific information.

A third area of prime importance to executives is the use of information services from outside the organization.

New computing services such as Motorola's Embarc actually give the executive the ability to wear a pager-like device or carry a small palmtop computer and receive industry-specific information as it occurs. This type of service, along with those that will follow, surely qualifies as a critical executive decision-making tool. Up-to-the-minute news flashes can keep an executive abreast of what's happening in a whole industry.

We believe that you can chart a course for executive literacy by starting with the very basics of keyboard knowledge and then learning electronic mail, learning query tools, and taking advantage of any of the newer information services that bring in, or should we say beam in, business information to the executive.

Specific knowledge and skills with word processing, spreadsheets, graphics, or database software are not quite as important for the executive. While it is certainly nice to be more or less familiar with these tools, it isn't as important as developing skills with query tools, e-mail, and online information services. Besides, it doesn't make good business sense for a highly paid executive to perform all his or her own spell checks or to spend lots of time dressing up fonts with a presentation graphics program. Ideally, the executive gets out the initial concepts and thoughts and then delegates the finishing touches to someone else (presumably, someone lower in skill and lower in pay).

Directors and High-Level Managers

Like executives, directors and high-level managers have often been a step removed from computer literacy. We believe this will change as more computer-literate people ascend the organizational ranks.

Specific skills needed by these individuals, as by the executives, include e-mail and query tools. We also recommend familiarity with constructing and manipulating spreadsheets, presentation graphics programs, and database programs.

Often, the work of managers involves consolidating information, reviewing it, interpreting it, communicating it, and making recommendations based on it. It is important that these individuals have enough hands-on skill to perform the "what if" or thinking process using information tools.

Other skills, such as word processing, use of electronic calendars and schedulers, and use of work-flow software, are important too. Many of these tools are remarkably easy to use and take little up-front learning time. Someone at the level of a manager or a director is advised to participate in whatever training is required to become comfortable and literate with these tools.

Business Specialists

The business specialist category includes the business professionals whose job it is to interpret data, make recommendations, or decide how the business should operate. They could be accountants, sales analysts, marketing specialists, or people with other titles. They frequently possess advanced educational degrees, such as MBAs.

In addition to all the skills required by executives and managers, business specialists should focus on learning database technology, ad hoc extraction tools, and report-writing tools that allow them to pull data out of databases. In some cases, it is helpful for these individuals to learn scripting languages or some kind of 4th-generation programming language.

Since individuals in these professions must interact with data on a daily basis, it is important for them to fully arm themselves with a variety of tools for the information-generation trade.

These professionals should also grow familiar with the use of communications software. Because they frequently interact with data from a variety of sources and people from a variety of groups, it is important that they can communicate on a number of levels. Communications software, which allows access to remote data sources or to public information forums, is an important asset.

User/Analysts

For our purpose, we will define user/analysts as people whose jobs require them to create or maintain information. These jobs may be titled analyst, specialist, executive or administrative assistant, or sometimes secretary or clerk.

People who fall into this broad category must learn many of the mechanical skills associated with office productivity and communications software. They are likely to use six to eight different software packages in the course of a normal work day. These software packages can include:

- E-mail
- Spreadsheets
- Word processors
- Presentation graphics programs
- Electronic calendar
- Database or filing software
- Work-flow or forms-based software
- Data access or query tools
- Report writing tools
- Data input software
- General communications software

Of course, the individual's specific job will dictate what tools are needed. Over the course of a career, a typical employee will work with nearly all these tools.

Blue- and Pink-Collar Workers

Another growing group of computer users includes blue- and pink-collar workers. These are people in clerical positions or who work on the shop floor, drive delivery trucks, or work in warehouses. As technology permeates the workplace, these occupations are not exempt from the need for some form of computer literacy.

Frequently, workers in this category don't need to learn all the nuances of office productivity software—many of them don't work in offices. But, they are often required to interact with company data by building up information about how company goods and services are being made or sold.

Blue- and pink-collar workers are often required to learn company-specific applications, such as the manufacturing system, the order processing system, or the inventory tracking system. Frequently, they are involved with the details of manufacturing processes or purchasing transactions. Their accurate input or involvement is critical to all the information that eventually follows.

As a result, it is important for these people to gain specific knowledge of core company computer applications. Training is generally handled as an on-the-job activity.

Over time, we believe, these people will also find it necessary to gain additional computer skills in the areas of e-mail and communications software. They may also be candidates for multimedia software for educational experience. Already, some companies are experimenting with making training in factory processes available to workers through multimedia sessions.

So, What Can and Should You Do Next?

We firmly believe that most jobs, as they evolve through the 1990s, will require more and more computer literacy. We strongly encourage workers in all professions to personally assess their own literacy and do what it takes to become comfortable with computers.

From our vantage point, we see the emergence of a class difference between information technology "haves" versus "have-nots." Moreover, as non-work-related information sources, like encyclopedias, newspapers, and social groups, become available and accessible by computer, there will be more advantages to computer literacy than just keeping your job.

Chapter 8

Fighting and Winning the Political Battles in an Era of Change

Have you ever suggested a new computing or communication idea only to hear something like "That will never work" or "We've done it this way for years, it works, why should we change now?" If so, you may have been exposed to people wallowing in a severe case of denial. As we discussed in Chapter 7, people can and do exhibit negative reactions to change. Changing out information technology platforms is enough of a change to make some people break out in hives. Don't worry, it's not contagious, but it can be detrimental to the health of your organization.

In this chapter, we'll talk about the challenges of change. We'll identify an insidious malady, which is sometimes called "mainframe myopia." It is a condition, normally found in large companies, where the information services people tend to grasp and hold all computing power. They don't want to relinquish control; they don't want to treat the business people as internal customers; and they don't want to change. Interestingly, "mainframe myopia" doesn't require a company to have a mainframe. In fact, it really isn't connected with a computing machine at all. Instead, it's a mind-set and can exist anywhere. We'll focus on how to spot the myopia and how to correct the problem before it chokes effective computing practices in the company.

Symptoms of Myopia

You may be asking what, in fact, is computing myopia and how do you spot it. More importantly, how do you cure it and wipe out the disease before it affects your entire company? The former problem is a lot easier than the latter, and before you can tackle the problem you have to recognize it exists.

Organization: Too Technology Focused

In large organizations, the easiest way to recognize myopia is to take a look at the Information Systems (IS) Department organizational chart. Does the organization mirror a closely controlled computer architecture, or does it look like the rest of the company? Is the IS department too technology focused? Is the NCP programmer king of the hill, or is there a Vice President of Internet Protocol (IP) addressing? Besides that, what is NCP anyway? Does everything revolve around 3174s, or around multi-protocol routers? If you picked the first half of each question, then, more than likely, your organization has the condition. The next question you have to answer is, how serious is it? Find out who got the last promotion. Was it another techno-nerd?

Methodology Madness

Often, IS departments follow strict methodologies for developing computer applications. Most of these methodologies were developed and popularized when the computing world was a very different place. When computing resources cost $250,000 per MIP (million instructions per second), it was prudent not to waste them. Besides, back then, people power was cheap, and so it made sense to optimize computer resources even at the expense of adding people to a process.

But today, things are very different. With the same computing firepower costing less than $500, it is no longer wise or prudent to conserve every dot and dash on a computer screen. Furthermore, the price of people power has escalated sharply. Even a relatively unskilled clerical job now costs three times the salary level it did only 15 years ago.

For purposes of illustration, in Table 6, we list the cost of delivering a unit of computing power (MIPS) to a corporate desktop since 1980.

TABLE 6 Cost of Delivering a Unit of Computing Power to a Corporate Desktop

Year	Cost
1980	$250,000
1985	$25,000
1990	$2,500
1993-4	$500

Clearly, with cheaper resources available, it is no longer wise to use methodologies that protect and preserve machine resources.

Backed-Up Backlogs

For a good gauge of how bad your organization's myopia is, map out the path a new requirement takes from the time a business person says "I need it" to the time the final product is delivered. Often, methodology-burdened, mainframe-centered organizations tend to centralize all project approval and development.

As a result, the "priests" in the glass house have to bless the project and then appoint a set of disciples to do the real work before any progress is made on meeting the needs of customers. If you're lucky, this process will take less than twelve months, but it is likely to take longer. The more prevalent scenario is that the business person or internal customer ends up waiting one to three years for a desired capability, or he pays an exorbitant fee to get it earlier.

Of course, this scenario assumes that the business department doesn't get disgusted, give up, or decide to strike out on its own and try to build the computing application itself. While these pioneering efforts do work, they can also cause problems. Often, they create disparate, unconnected workgroup systems, and organizations have to wrestle to incorporate these smaller focused systems into the needs of an enterprise-wide information infrastructure.

Poured-Concrete Cultures

Embedded "concrete-like" culture is another pressure point for a myopic organization. In many instances, organizational culture tends to take on a life of its own. Bureaucrats who have grown up in a given culture tend to be stuck in the box and have a hard time peeking over the side. In a June 28, 1990, issue of EDN, Richard A. Quinnel wrote a rather humorous column entitled "Llama Alert!" illustrating this point. The following is an excerpt from his article.

> In the early 1940's, so the story goes, the Army wanted a dependable supply of llama dung, as required by specifications for treating the leather used in airplane seats. Submarine attacks made shipping from South America unreliable, so the Army attempted to establish a herd of llamas in New Jersey. Only after the attempt failed did anyone question the specification. Subsequent research revealed that the U.S. Army had copied a British Army specification dating back to Great Britain's era of colonial expansion. The original specification applied to saddle leather. Great Britain's pressing need for cavalry to patrol its many colonies meant bringing together raw recruits, untrained horses and new saddles. The leather smell made horses skittish and unmanageable. Treating the saddle leather with llama dung imparted an odor that calmed the horses. The treatment, therefore, became part of the leather's specification, which remained unchanged for a century.
>
> The moral of this story is that too many people get stuck in the "we've always done it that way" mode and thus become slaves to a process regardless of the rationale and necessity behind it. Do you have people in your organization who fit this mold?

Customer Customs

Another symptom, and probably the most important to address, is customer dissatisfaction. In organizations that have myopia, you will probably find a lot of unhappy IS customers. While the incumbents of the IS shop seem to be very content with life and persistently pat themselves on the back for jobs well done, their customers are getting restless.

Sadly, their customers probably have stopped complaining, because it never really did any good. Instead of wasting their breath, the customers went out and educated themselves in PC and LAN technology. This in itself, however, isn't all bad. The fact of the matter is that these "home-brewed" solutions, while rather crude, have forced IS staff to take distributed computing seriously.

For example, we just visited an organization whose entire LAN support contingent in the IS shop consisted of a single person, while the mainframe staff numbered well over 100. What had happened over the years is that the local hacker within a division became the office automation expert. When the line of business managers realized that there is more to life than sending e-mail to Bob in the next cubicle, they all turned to IS to provide an interconnectivity solution. Many IS managers just gave them a blank stare in return.

Creating an Environment for Change

The bottom line is that, whether you like it or not, your organization will have to make the transition from being mainframe-centered to being network-centered, information-centered, and, finally, people-centered.

If you are one of those astute individuals who has already recognized this, then the rest of the chapter will outline some basic strategies you can use to pursue, push, shove, go around, or drag your organization into the twenty-first century. Depending on who you are (and how you became responsible for causing change), you'll have a number of strategies you can pursue.

There are several options to consider when trying to overcome computing myopia within your organization. All are situation dependent, and you may find yourself using a combination of them in order to accomplish a particular project. The different strategies available to you include evangelist, authoritarian, stealth fighter, guerrilla, and big brother.

Before you decide which strategy or combination of strategies will work best for you, you must consider several factors that will have an impact on your choice. The first is the level of influence (and authority) you have within your organization. The second is the level of influence that your opponents have within the organization. The third is the criticality of the project or function. The last factor affecting choice of strategies is the fear of change or risk factor.

Again, keep in mind that each case will be different, and you will constantly need to adjust your strategy. As we discuss each of the individual strategies, consider how you would apply it in your daily work life. Pick a project that seems to be stuck in a logjam and picture how you would utilize these strategies to set the project free.

Energize as an Evangelist

Evangelism is the role of the "cheerleader of change." It is also the one underlying strategy that does cut across strategy options. While almost anyone, at any job level, can assume the role, it is one that requires a full commitment. It is hard to be a part-time evangelist. In fact, it is impossible to evangelize on an occasional basis. Being an evangelist means holding onto a dream and being passionate about achieving that dream. In this case, "dream" is really synonymous with "vision." It's not easy to sell a vision. You can't force anyone to buy into a vision; rather, they have to be converted. "My experience tells me that an energizing, inspiring vision is the key to mobilizing support. This vision is the picture that drives all action. It includes both deeply felt values and a picture of the organization's strategic focus." (James Belasco, *Teaching the Elephant to Dance.*)

An evangelist is a pied piper. He or she must play the music before anyone will follow. Live the vision, and tie every project back into achieving that vision. It is amazing how the vision will spread if someone keeps talking about it and living it. Correspondingly, it will die if someone stops making noise.

There is a group within the Air Force Space Systems Division that proved vision doesn't always spread from the top down. This group's role in life is to provide communications support to launch operations. Normally, this only entails providing some basic telephone service. However, the captain in charge of the group recognized that, through the use of high-speed communications, they could reinvent the launch process. His vision was simple: "Let's leave all the test equipment and test engineers at the factory and provide the necessary connections with high-speed communications instead of loading several cargo planes full of equipment and sending 200+ people down to the launch location for six months."

He was so sure his idea would work, he set off on a daily quest to sell the vision to all levels within the organization, including the IS staff responsible for building the network. Slowly, the communications infrastructure was built. In the initial stages, some brave customers used a few low-speed circuits. As the network's reliability was proven, the speeds and number of the circuits began to increase dramatically. By this time, the customer had totally incorporated the vision. When the last phase of the captain's vision is finally implemented, the Air Force will end up saving several million dollars per launch.

As our captain illustrated, selling the vision at all levels is important. Some key points to keep in mind about vision: Vision solely maintained at the top never gets internalized by the people who have to make it happen; vision solely maintained in the trenches never really gets the resources to achieve it; vision solely maintained in the IS shop never gets incorporated into the operational environment; and vision solely maintained by the customer results in a former customer.

Move Aggressively as an Authoritarian

The authoritarian strategy requires a strong leader and a person in a position to force things to happen. This individual can direct projects, and the projects directed tend to move from the top down. The leader in this case must consider himself an agent of change and be willing to take a few risks to achieve a greater reward.

I'm sure we have all heard, "You don't get fired for buying IBM." At least that used to be the case. Not so anymore, and, as we would argue, that attitude has forced many Chief Information Officers (CIOs) into the unemployment line. If you are this type of leader or you know of one within the organization, then you can interject one of the primary ingredients toward eliminating myopia: CHAOS. Since myopic individuals remain in a constant state of denial, they can not handle change. Eventually, many will be overcome by events and step aside to let the new guard step up to the challenge.

The author encountered such a situation years ago. The department manager took the authoritarian approach to force the organization to adopt a client/server architecture. Everyone realized that the entire computing environment needed to be upgraded. No one, however, really could agree upon which direction to go. The manager stepped in, made a decision, set some goals, and then allowed the smart "young" troops (some of whom were not at all young) to take off and implement them. While they were in the process of pulling the solution together, the manager and members of his senior staff set off to interact with customers and "prep" them for what was ahead. The result: 1000+ desktop computers were converted to state-of-the-art workstations, and three mainframes were shut down in a span of eight months.

Snag a Sponsor

A closely related strategy to the authoritarian approach is to find a "big brother" or a sponsor from somewhere in the business units or senior management. In other words, a good way to put the customer first is to find someone to become the champion of computing change. It works best if this champion is a customer (aka business person) himself or herself.

This will drive the myopic old guard crazy, since they can't flat out tell the customer to take a hike and they have few resources to fight a high-ranking customer.

In reality, your sponsor won't have to do a whole lot. He or she will just have to show up for important meetings, nod approvingly at progress, remove occasional roadblocks, and make sure that the project has the appropriate amount of cash to succeed. Maintaining and portraying a business focus is the key.

Get a Few Guerrillas

If you are not lucky enough to snag a sponsor, or you don't have a boss who has totally bought into your vision, there are still options for action. In these cases, you may have to resort to covert tactics in order to successfully attain your vision. One such tactic is to apply the principle of guerrilla warfare.

The basic principle of guerrilla warfare is to pick and choose your encounter with the enemy carefully. Professional warfare strategists look for the moment when they have an overwhelming force capable of achieving a tactical victory, even though overall the enemy is far superior. When not engaged in battle, forces simply blend in with the surrounding environment.

The North Vietnamese and the Angolans successfully used this technique to achieve their objectives. Although you won't be using guns and bullets, the battle to overcome computing myopia can be just as intense.

If you must grow your own guerrilla group, here's a plan of attack. Your first objective is to search out and find a few small projects that the glass house considers to be trivial. Use these projects to achieve some tactical wins with business users. As your successes accumulate, start searching for larger projects to tackle.

Eventually, these small wins will become noticed. You will wake the giant, and he will try to crush you. Speed, flexibility, and wit will keep you from being pushed back into the corner. It is at this moment that you fall back on your tactical victories to illustrate that your approach does work. If you don't succeed in convincing an authoritarian figure or customer sponsor, then revert to the basic principle and look for projects where you do have a tactical advantage. Sooner or later, you will find a killer application that will force the organization to the bargaining table and commit it to mainstreaming the technology.

An IS shop for a large systems integration firm used this strategy to successfully incorporate their network management software into division proposals for external customers. The software started as an in-house effort to help meet the demands of managing a worldwide computer and communications network. The product eventually evolved to where it became a set of generic tools that could be applied to a variety of tasks. During the life of the project, all of the organization's divisions had the opportunity to invest in the development in order to enhance the product.

None did, and eventually the IS shop looked outside the organization for a strategic partner. They found one, and when they did the rest of their organization finally realized that the product had commercial market potential. Now, all the divisions have lined up in hopes of becoming the internal corporate sponsor for the software.

Steal a Victory with Stealth Techniques

The final strategy to consider for overcoming myopia is a stealth strategy. Just as the name suggests, this strategy requires concealment. If myopia or any other resistance to change is seriously entrenched within your organization, then this may be the only strategy that will work.

When you take on a stealth role, it's best to follow Admiral Grace Hopper's advice: "It's easier to ask for forgiveness than to ask for permission." In other words, just do it. I'm sure you're sitting there saying that's easier said than done. However, you may be surprised at the various avenues that are probably already available within your organization. The first would be a suggestion program or innovation center, the latter being the more progressive of the two. If your organization takes employee comments seriously, then use that avenue. The "idea investigator" will usually be someone from outside your immediate office who is open to some new ideas. If the process works, you will get a champion appointed who will help guide your idea through the bureaucratic maze.

A second alternative is to work through an internal customer. This is different from finding a customer sponsor. Rather, the customer fronts the entire project, and you just become a temporary member of his staff. This approach will take the project completely out of the hands of the IS shop. More than likely, the IS management will fight it or at least complain that IS should have a deeper involvement. It will take a strong customer to just say no.

A final stealth strategy is to bring in an "expert." Find an industry expert who basically agrees with your vision and ask him or her to come spend some time with your organization. As the old adage goes, an expert is nothing more than a stranger with a briefcase wearing a nice suit. You'll find that people will tend to give more credibility to someone they don't know than to someone they work with on a day-to-day basis.

Change Is Chaotic

As you continue to come up against myopic attitudes, which truly translate to resistance to change, you will find that it can be frustrating, exhilarating, and confusing all at once. At best, the environment is simply chaotic.

It is important that you remember that everyone is trying to do a good job, regardless of individual motives and views of how to get that job done. It's human nature to be afraid of the unknown.

Often, change-resistant people are simply people who have gotten comfortable with the status quo and feel good in familiar surroundings. Fear is the biggest deterrent to change.

As an agent of change, you will find that your basic challenge is first to conquer that fear within yourself and then to help alleviate it within others. The more threatening you appear to another individual, the more resistance you are going to encounter. Everyone has a role to play and a contribution to make. Highlight and accentuate that aspect, and you will be able to truly conquer mainframe myopia.

Chapter 9

Reengineering

Perhaps the hottest trend in corporate America today is *reengineering.* Management guru Michael Hammer of the Center for Reengineering Leadership defines business reengineering as:

> ...the fundamental rethinking and radical redesign of an entire "business system"—the business processes, jobs, organizational structures, management systems, and values and beliefs—to achieve dramatic improvements in critical measures of performance.

Many pundits believe that reengineering is the salvation of American industry. It is certainly getting its fair share of attention in the boardroom and throughout the executive ranks. The consulting group CSC Index, Inc. recently conducted a survey of 407 businesses and found that engineer72 percent of the companies are reengineering their business processes. The survey notes that top-ranking IS executives, both here and abroad, list "reengineering business processes" as their most pressing concern for 1993.

As a result of the worldwide recession and tougher competition, businesses must make radical changes just to survive. Those that don't are on the path to extinction.

In this chapter, we'll look at who is reengineering and why, how information systems affect and are affected by reengineering, and what you can do to prepare for and support your organization's reengineering efforts.

Who, What, When, Why, and How?

Reengineering is more than a passing fad; it is a survival tactic that virtually all large organizations are initiating or are already involved in. Corporations, government agencies, and educational institutions that have become burdened with inefficient and costly work flows are "reinventing" themselves. They are reevaluating the fundamental questions about their existence and missions.

Why Reengineer?

The question of "why" companies are reengineering their business processes is a simple one to answer: to survive. Of course, the more common answers are "to become more efficient," "to save money," "to better serve the customer," "to be more competitive," and so on. But, isn't it true that if a company does not do all of those things today it cannot expect to survive? The fact of the matter is that reengineering is not an option, it is a necessity for many organizations.

While corporations tend to focus more on the competitive benefits of reengineering, government agencies and educational institutions look for efficiencies and cost savings. Frequently, reengineering is a response to financial pressures. Certainly, the country's economic recession has been a factor in organizations' search for better ways to conduct business.

Organizations of all types are always searching for business nirvana: high productivity, low overhead, satisfied customers, repeat business, and happy owners or investors. In years past, corporations focused on back-office automation to help reach these goals. The trouble is, we automated the *old* way of doing things. If a manual process was already inefficient, automating that process didn't really help. In essence, we paved the cow paths instead of creating highways. Moreover, we have automated just about all that we can. It's time to look for other ways to achieve increased productivity.

Who Is Reengineering?

So who is doing it? Just about everyone. *Computerworld Magazine* recently conducted a survey of 102 top information systems executives. The survey revealed that reengineering and its cousin, total quality management, are key activities in supporting the representative companies' business strategies.

We could fill this book with examples of organizations that have undertaken a reengineering effort. Here are just a few of the cases that we are familiar with:

Richmond Savings Credit Union, based in Richmond, British Columbia, Canada, redesigned its banking system to accommodate more services for its growing customer base. Instead of looking at the banking process as just a series of transactions—deposits, withdrawals, loans, and the like—the Richmond Savings team looked at the whole picture. They saw the value in integrating and making accessible *all* of the information about each client.

The reengineering effort was championed by the executive ranks and involved everyone, right down to the bank teller. The result has been the implementation of relationship banking, in which a customer service representative (teller) helps a customer manage his or her overall financial portfolio. Previously, the customer's accounts for savings, checking, and loans were all separate. Now, all of the systems are integrated to provide a total picture of the customer's financial standing.

At the heart of this reengineering effort is a new microcomputer-based information network, which provides the flexibility to respond to new situations almost immediately. Richmond Savings can add new products and services to its business repertoire almost as fast as they are created. Since instituting its process changes and implementing the new computer system in 1987, Richmond Savings has had an 850 percent increase in operating income, a 152 percent increase in assets, a doubling of accounts under management, a vast reduction of loan losses, a doubling of branch locations, and a reduction in operating costs for computer systems.

Ohio-based Banc One Corporation, the 12th largest bank in the U.S., also underwent a business transformation in recent years. Seventeen disjointed information systems were integrated to give bankers a complete profile of their customers. The bank now enjoys more cross-selling, highly targeted marketing campaigns and a speedy roll-out of new services tailored to individual customers.

Mutual Benefit Life has reengineered its processing of insurance applications. The old process typically took 5 to 25 days, while paperwork was handled by as many as 30 people in five different departments. The actual time that the application was being processed was minute in comparison to the time the application spent just being routed around.

At the company president's urging, radical changes were implemented at Mutual Benefit Life. Existing job definitions and departmental boundaries were nuked, and the new position of "case manager" was created. The case manager became responsible for the application throughout its life cycle, eliminating the old handoff routine and greatly reducing the time to process an application. Moreover, the new work structure more than doubled the volume of applications the company can handle.

And the list of examples goes on and on: Eastman Kodak Company, Cigna Corporation, Tenneco Gas, Bell Atlantic Corporation, Ford Motor Company, Hewlett-Packard, and more. Chances are that it won't be long before your organization joins the ranks of those that are reinventing themselves (if it isn't already).

How Should Reengineering Be Tackled?

As you might already surmise, reengineering takes a huge commitment from everyone throughout an organization, most of all from the executive and management ranks. The company leadership must set the vision for where the company is going and how it is to get there. Moreover, management commitment is required in order to get employees to accept the high level of change that will result.

Most companies assign a team to the task of reengineering. The team includes people—employees and outsiders such as suppliers—that represent all areas affected by the anticipated changes. The team can then look at the entire business from a cross-functional point of view.

The team should look at how a business process is conducted—for instance, how an order for product is handled, from placement of the order to collection of payment. Members should focus on the desired results, not on the current process. They must challenge the status quo and be willing to break "the rules."

The most important questions the team should ask are "Why?" and "What if?" *Why* must the order form be in triplicate? *Why* does the New York office process all orders for the East Coast? *Why* does Mary handle all requests for quantities of 100 or less? *What if* the order form were on line? *What if* the district offices processed their own orders? *What if* we didn't distinguish small orders from large ones?

The "whys" and "what ifs" help to separate what is fundamental from what is superficial. The answers are often based on implicit rules that are no longer appropriate. Most often, we find that our business processes weren't designed—they just evolved. And, the old processes often stem from a time when labor was cheap and computers were expensive, a state of affairs that no longer applies today.

It's easy to look at what's wrong with our present processes. It's not quite so easy to design new processes. This is where robust computer technology comes into play. The more flexible the technology that is implemented, the easier it is to change and improve the processes as business requirements change.

Where Information Technology Fits In

Information technology plays an interesting role in a company's business processes. Many systems, particularly those based on centralized mainframe technology, were originally put in place to automate a manual process. Over the years, the inflexibility of the systems themselves often came to define a process. In other words, a process evolved because that's how the computer could handle it.

We strongly believe that technology should be the *enabler* of a process, not the driving force. Your information systems shouldn't constrain what the company can do, when it wants to do it. Flexibility is the watchword.

Most experts on reengineering agree that a company gains the most efficiencies when the people who gather and process information also make decisions about it. This is what empowerment is all about. And computer technology can enable empowerment.

Open systems based on networking technology offer the most options for gathering, processing, analyzing, and distributing the corporate information base. When applied appropriately, computer networks offer a vast array of possibilities for improving work flow and empowering employees.

Desktop computing built around distributed networks can give people access to different kinds of information. For instance, a customer account representative could look up inventory levels, pricing information, credit limits, and past payment history before placing a customer's order. There is no need to hand off the information requests to different departments if the account representative can access it all herself. Just as Richmond Savings Credit Union learned, integrated information can lead to increased sales.

Computer technology in the form of expert systems can help employees with limited experience make sound decisions. Again, this comes back to empowering employees to make their own decisions and reduce the time wasted when handing off responsibility.

Technology in the form of work flow systems and imaging can reduce bureaucracy and paperwork. E-mail, electronic signatures, routing software, and decision support systems speed the delivery of information. Applied appropriately, today's computer technology can eliminate wasteful activities and increase overall productivity. Recall our example of the travel request form in Chapter One. By redesigning the arcane process and using e-mail to initiate the request, we save both time and money, and isn't that what it's all about?

So, we maintain that technology enables new processes. But, what happens to the old information systems when a new process is designed? We believe they should die a merciful death. There is no room for inefficient legacy systems in the new business model. Remember the word "radical" in our definition of reengineering? Radical change is necessary for radical improvement, and that change should include the company's information systems.

Fortunately, the plummeting prices of computer technology, particularly in the microcomputer arena, soften the age-old barrier to implementation: cost justification. Regardless of the price, enlightened business executives understand the capabilities of new computer technology and realize that this technology has to be applied in totally new ways in order to position the company as a leader in its business sector.

What Can You Do to Prepare for Reengineering?

Whether you are a corporate executive, an information systems professional, a business line professional, or anything in between, you should prepare to support your organization in a reengineering effort. The chances are that before too long, your company will get caught up in the groundswell of reengineering activity.

Get the Broad Picture

Perhaps the most important thing you can do is to start looking at the broad picture of your business. Look beyond the niche of your particular job, and view the business as a whole. Get to know who does what and the relationships among the various departments. Once you know who the players are, it will be easier to understand their jobs and work processes.

Don't stop your "fact-finding mission" at the company walls. Get to know your customers and suppliers. If (when) your company does take on a reengineering effort, it is likely that some of these outsiders could be involved or affected. For instance, if the company revamps the way invoices are paid, your suppliers will surely be affected.

You may want to include them on your reengineering team in order to get their fullest participation later, during implementation of a revamped process. Moreover, it's quite possible that these folks will have some very good suggestions on how to improve your business, since they can look at the current methods with an objective eye. Since they have no emotional tie to the current way of doing things, they can help eliminate the frivolous activities.

Also, take a look at what your competitors are doing. While it won't be easy to learn exactly how they are doing things, you should still try to gather some general information. A covert espionage campaign won't be necessary; simply take advantage of your business contacts. When you go to trade shows and conferences for your industry, talk to other people about their business. Surprisingly, you will find many employees of competing companies quite willing to talk openly about their business. You can learn a lot during a casual conversation.

Another way to learn about the competitors is to subtly ask vendors and suppliers about them. Chances are, your suppliers will be their suppliers too. While you probably won't get the detailed story, you may learn about general strategies. For instance, a computer sales representative might tell you that your competitor is buying Brand X hardware and Brand Y software for a new order placement system.

Studying the broad picture of your company and the forces around it will put you in a good position to contribute valuable ideas in a reengineering effort.

Get Acquainted with the Latest Technology

These days, it's almost impossible to separate a company's work processes from its applied computer technology; one depends on the other. When your company reengineers its work processes, it will surely need to update the corporate information systems.

You can prepare for the inevitable by getting acquainted with the latest tools and technology. Your job within the company will determine just how much you need to know. Obviously, an IS professional will need to know much more about computer technology than a business line professional will. If you aren't an IS professional, however, don't feel intimidated by the technology or exempt from learning. No corporate professional should be excluded from technology decisions that will affect his job. Moreover, a higher level of computer literacy will increase your value to the organization. Strive to be a Frank Flash instead of a Joe Average!

There are a lot of choices in the computer technology arena; it may be hard to decide how to spend your study time. We advise that you concentrate your efforts on networking solutions based on client/server technology. The importance of desktop computing will continue to grow throughout the decade. Focus on the tools and techniques that will allow non-IS employees to gather and analyze data for themselves. Data query tools are especially important.

Integration of disparate data systems will also grow in significance. Focus on tools and products that will bridge the gaps among your information systems. If your company can build intelligent interfaces among systems, it may be able to avoid (or at least delay) the costly redevelopment of existing systems.

Flexibility is an important element to consider. That's why we feel open network computing is so fundamental. Your company will be able to respond to new business requirements much more quickly in an open environment. We urge you to avoid proprietary systems like the plague!

As you look at the current and emerging technology, don't overlook the value of the human interface, that is, how real people will interact with their computers. Today's leading products are based on graphical user interfaces (GUIs). Emerging products are based on handwriting (pen-based systems), touch screens, and even voice recognition. It is worth the time now to evaluate the potential of the technology, even if it may be years before some of these newer technologies are stable enough to build your systems around.

Obviously, there is a lot to know about computer technology, and no one person can become an expert on everything. But, the more you are familiar with what is available, the more valuable you can be to the reengineering team that has to decide *how* to implement the new systems.

Develop a Questioning Attitude

As we said, the important questions in a reengineering effort are "Why?" and "What if?" There is no need to wait until a formal reengineering venture is under way before you begin asking those questions.

Start with your own activities. Take an objective look at what you do each day. Have the actual tasks of your job become more important than the outcome? Have you lost sight of the desired results? Do you do things a particular way simply because that's how they have always been done? Begin a reengineering effort for your own job. Who knows, it just might snowball to others around you and lead to a widespread endeavor.

Remember that results are what is important. There are any number of ways to arrive at a goal; you want to find the most efficient and effective way to get there.

Preparing for Change

Reengineering affects a lot more than just a work process or an information system. Job assignments come and go, and personal responsibilities and accountability change with them. Career paths and recruiting change with the ebb and flow of job assignments. You or your associates may need to acquire or develop new skills and knowledge. Even compensation policies may need an overhaul.

Management structures are frequently shaken up when work processes are redesigned, usually to a flatter organization. Business relationships change, we hope, for the better. Even the corporate culture may change as the status quo is tossed out the door.

Does all this change sound like chaos to you? It is, at least until a new set of routines emerges from the ashes of the old style of business. If you are steadfast and resourceful, the phoenix that does emerge will be streamlined, flexible, and ready to do battle once again.

Reengineering is a radical tactic to get your entire company back to the basics. In Chapters 10, 11 and 12, we'll take a look at a few other tactics that are used primarily to shake up corporate information systems.

Part II

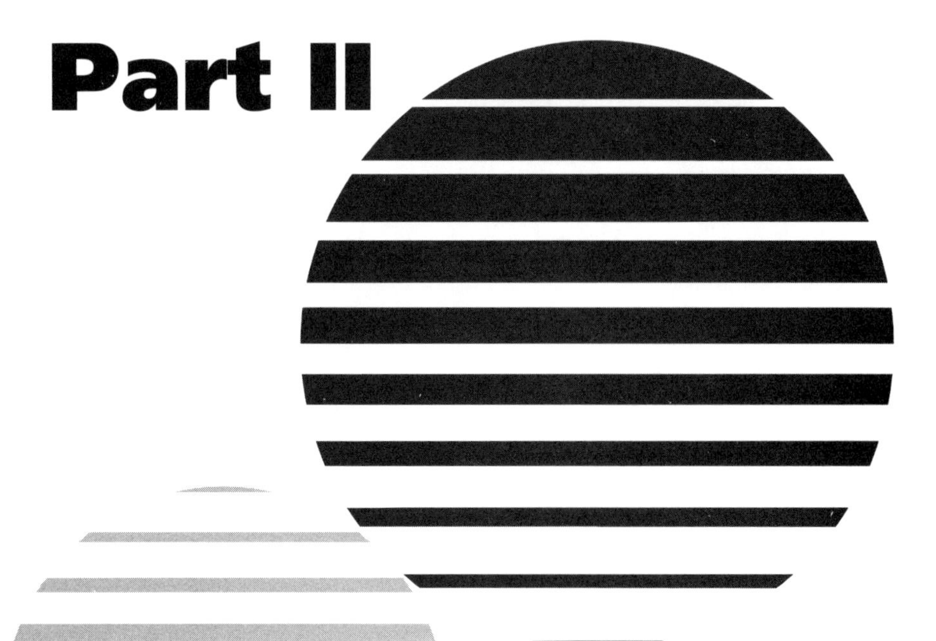

Information Weaponry: The Strategies and How They Affect You

Chapter 10

Outsourcing

Outsourcing refers to turning over the operation of all or part of your IS responsibilities to an outside agency. It isn't new; in fact, outsourcing has been around since the early days of external payroll processing. What is new is the extent to which companies are outsourcing their IS operations today. We see deals that run the gamut from contracting out simple PC support all the way to handing over the entire IS operation to someone else.

Max Hopper, senior vice president of information systems at AMR Corporation, says, "outsourcing of some kind is a foregone conclusion for most IS operations." He says the question is not *whether* to outsource, but *how much* to outsource, and *to whom.*

Outsourcing, also known as facilities management, represents a panacea to many harried IS executives. They can take the pressures of tight budgets and growing demand for services and dump them on a third-party vendor that is contracted to provide a specified level of service and performance at a fixed price. Executives also like the idea of paying for IS services on a per-use basis, rather than as a constant expense.

Outsourcing can be a blessing or a curse, and sometimes both. In this chapter, we'll take a look at the upside and downside of this popular coping mechanism. We'll look at how it can be applied and share a few tips for success.

The Reasons for Outsourcing

For almost all companies, the primary reason for outsourcing is economic. Studies show that savings can run as high as 15 to 40 percent, largely through economies of scale and the ability to leverage expertise and resources.

New York-based First Manhattan Consulting Group conducted a study of outsourcing in the banking industry and found that outsourcing typically saves 40 percent over in-house IS support. Even if the internal group takes steps to improve its own operations, savings usually reach around 20 percent. This margin gives the outsourcer a significant advantage—enough to pass savings along to the customer and still make a profit.

In a recent Computerworld Magazine survey, IS executives, maintain that "saving money" is the biggest benefit of an outsourcing deal. These executives believe they can get more done with a lower financial investment. Moreover, they anticipate getting greater expertise from an outside source.

For some organizations, cost *containment* is more important than cost savings. With trimmed IS budgets and escalating costs, IS managers feel that outsourcing is a good way to hold the line on expenses.

Economics aren't the only reason why many companies are turning to outsourcing. The market research firm Ledgeway/Dataquest cites some other common reasons: the expectation of improved performance, reduced management in the IS department, shortened implementation time, and the acquisition of expertise. Some companies are simply looking to return to their roots—to get out of the data processing business and back to the basics.

Eastman Kodak started the ball rolling in 1989 when it turned its entire IS and telecommunications operations over to outsiders. Kathy Hudson, Kodak's director of IS, justified the move quite simply by saying, "IBM is in the DP business and Kodak isn't. IBM runs our computer center as it's supposed to be run—as a profit center rather than a cost center."

Lane Jorgensen, manager of market planning and support for Integrated Systems Solutions Co. (ISSC), the outsourcing arm of IBM that has the Kodak contract, says that there are two forces driving the outsourcing market: global competition and asset utilization. CFOs are scrutinizing the corporate assets and deciding where they should be spending money. In Kodak's case, the choice boiled down to putting capital into new film and camera products or into information systems.

Government agencies often have very different reasons for outsourcing. The Information Technology Association of America reports that professional staff shortages top the government's list of reasons to outsource. Ceilings for both head count and wages prevent agencies from directly hiring the people they need. Outsourcing provides a convenient way to get around the regulations.

Bill Dvoranchik of the government services business unit of EDS Corp. says that "outsourcing also appeals to the risk-averse federal IS manager who can take credit when it works but blame a contractor when it does not."

Who's Outsourcing?

Back in 1989, Kodak was the largest of the Fortune 500 companies to place its faith (and fate) in the hands of outsiders. Kodak's move unleashed a torrent of activity from other companies, making the early 1990s a boom time for outsourcing megadeals. On the heels of the Kodak deal, a number of companies sought to save money through outsourcing, as illustrated in Table 7.

TABLE 7 Companies That Have Outsourced IS Operations

Company	Contract Value	Duration	Vendor
Del Monte Foods	$150 million	10 years	Electronic Data Systems (EDS)
Health Dimensions, Inc.	$20 million	5 years	Integrated Systems Solutions Corp. (ISSC)
J. P. Morgan and Company, Inc.	$20 million	5 years	BT North America Inc.
Signetics Corp.	$100 million	10 years	Electronic Data Systems
U. S. Dept. of Housing and Urban Development	$526 million	12 years	Martin Marietta Corp.
General Dynamics Corp.	$3 billion	10 years	Computer Sciences Corporation (CSC)

The table lists just a few of the recent agreements. Corporations and government agencies all around the world are turning to outsourcing as a management tactic. Ledgeway/Dataquest predicts that the worldwide market for outsourcing services will reach nearly $65 billion in 1995, up from $28.9 billion in 1990.

Government contracts currently make up the second-largest outsourcing market, with an emphasis on processing operations, software maintenance, and application software. Ledgeway sees the top ten industries for outsourcing by 1995 as discrete manufacturing, state government, distribution, commercial banking, diverse services, federal government, telecommunications, process manufacturing, health care, and insurance.

What Can Be Outsourced?

Outsourcing comes in two flavors: the full farming-out of all IS operations, and the selective use of vendors for particular tasks. While both approaches are popular, selective outsourcing is far more common. Many companies choose to contract just a few troublesome areas, such as application development or network operations. Selective outsourcing has also gained momentum from the recent trend to shatter the glass house and move networks, databases, and application development out into the user community.

Private data networks are a prime target for outsourcing, with global communications carriers stepping in to do thc job. Network outsourcing is a $2.3 billion industry that is growing by more than 20 percent per year. With telecommunications becoming more complex, IS managers are increasingly looking to outsiders to help manage corporate communications services.

Telecommunications services companies can typically offer router, bridge, and LAN interconnection services, in addition to network design and support services and PBX management services. This breadth of services places companies like AT&T, MCI Communications Corp., US Sprint Communications Company, and Infonet Services Corp. in direct competition with the traditional outsourcing vendors like EDS, CSC, IBM, and Anderson Consulting. All of this competition bodes well for the corporate customer, who can pit one vendor against the other during the contract bidding phase.

Application development is another popular IS task that can be outsourced. Howard Anderson, president of the Yankee Group, cites three conditions when it is reasonable to outsource development.

1. When a company has a tremendous development backlog
2. When the development challenge looms large and the company doesn't have the staff or skills to address the problem adequately
3. When a firm has a large, one-time project

There are two major risks to outsourcing strategic applications: breach of confidentiality and loss of timeliness. If an application is truly strategic to a company's operations, the organization cannot afford to risk making it public. Even confidentiality agreements may not be enough to protect the company's application assets. Timeliness can be another big concern. Will the outsourcer meet the scheduled delivery date? Are the resources available to meet the goal?

PC and workstation support is another common target for outsourcing. Third-party vendors are best able to realize economies of scale in activities like PC burn-in and configuration, software installation, troubleshooting, and repair.

While almost any IS activity can be outsourced, some lend themselves more to this treatment than others. Rod Canion, chairman of the IS consultancy Insource Management Group, Inc., suggests that companies outsource only "commodity functions." Canion believes that IS support of strategic services should be kept in house.

The Upside to Outsourcing

Outsourcing has many good and bad points to it. Moreover, what one company views as a positive point another may see as negative. As with any other IS management tactic, it all depends on which side of the fence you're sitting on. In general, however, companies that have outsourced at least some portion of IS operations have identified the benefits listed below.

Cost Savings

As we've already mentioned, saving money is the top reason why companies outsource their IS services. For most of the long term megadeals, it's too early to tell if, in fact, the companies are realizing a reduction in overall costs. Most companies, however, conduct detailed analyses before signing the contract and project significant cost savings over the years.

As much as we talk about anticipated cost savings, outsourcing doesn't guarantee them. In fact, outsourcing can sometimes cost more than doing the work in house, but it frees valuable internal resources to work on critical systems.

For organizations that do realize cost savings, they usually come from the economy of scale that the vendor provides. For example, most large companies keep at least one communications expert on the IS staff. If an outsourcing vendor has ten clients, it doesn't need to have ten communications specialists on the payroll. Perhaps only four or five are needed to accommodate all the clients.

Economy of scale can apply to equipment as well as people. It's quite common for the client to use a portion of the vendor's computing resources, as in a timesharing arrangement. In fact, in some outsourcing agreements, the vendor actually assumes ownership of the client's equipment. In its deal with CSC, General Dynamics sold its computer facilities as well as the computers. Larry Feuerstein, vice president of GD's information systems, says, "We don't have a capital budget for computing any more. We can spend it now on the core business and invest in plant equipment to build products."

Business Focus

Like Kodak, many companies that have outsourced report that they are now able to focus on what's really important—the business. In this case, the outsourced IS operations are usually seen as a utility function rather than a strategic function.

With the day-to-day IS activities being managed and performed by someone else, the key strategists in the IS department are free to concentrate on the core business.

Expertise

Vendors bring to the table something that isn't always easy to find: a ready talent pool with very specific expertise. This is particularly helpful when a certain skill is needed only for a short time. For example, a company may need a network engineer to design the layout of the corporate network. Once the design is completed, the engineer is no longer needed on a full-time basis. The vendor can reassign this person to another client's project.

Outsourcers tend to have a lot of IS veterans at their disposal. Thus, they make a broad spectrum of experience available to the client company—experience that may be too time consuming or expensive to acquire otherwise.

Outsourcing giant Computer Sciences Corporation keeps a detailed database on the skills of all its employees. When a particular skill is needed for a client, CSC can search its database and find just the right employee for the job.

Staffing Options

When IS functions are outsourced, the internal IS department can be reduced or redeployed. From management's perspective, this staff reduction results in cost savings. Of course, that's no consolation if you're one of the employees on the layoff list.

Most companies negotiate a "home" for their employees on the vendor's staff. For example, when General Dynamics signed its $3 billion agreement with Computer Sciences Corporation, 2600 of GD's IS employees were transferred to CSC. This is a win-win situation for everyone. GD saves money on salaries and benefits, CSC gets a knowledgeable staff, and the GD-turned-CSC employees still have a job. As far as the employees are concerned, the only change is which company logo appears on the paycheck each week.

Not everyone is trying to reduce staff. As we've already seen, government agencies use outsourcing to *gain* staff.

The ability to do resource leveling is another benefit to outsourcing. Many companies find it is easier to rely on a vendor than to increase or decrease staff on short notice. For government agencies that deal with complex employment regulations, this fact can be essential.

Eliminating Drudge Work

Activities that are considered "drudge work" make good tasks to outsource. For instance, keying data from paper sources. (Better yet, you should re-engineer the process to capture the data electronically in the first place and eliminate the paper.) Nevertheless, there will always be some activities that are necessary yet less than exciting. Rather than use your own valuable resources for these tasks, let a vendor deal with them.

Tax Relief

At the present time, outsourcing fees are deductible as a current business expense, bringing companies some much-needed relief at tax time. Of course, this is subject to change over the life of a long-term outsourcing contract, but at least companies can be enjoying the benefit today.

The Downside to Outsourcing

Just when you thought outsourcing might be the answer to your prayers, we need to take a look at the downside of this approach. There are companies that have been involved in outsourcing contracts long enough to tell us what went wrong with their agreements. Most often, the problems stem from conflicts over items that were not spelled out in the contract. Some companies have been so unhappy with outsourcing that the contracts needed to be cancelled completely, often at great expense. We share the benefit of their wisdom.

The Nickel-and-Dime Syndrome

It's impossible to anticipate every service and quantity of service that will be required over a long period. It's inevitable that the client will want services additional to those that were specified in the original contract. This fact leaves the door open for what we call "the nickel-and-dime syndrome."

The root cause of this syndrome is "asking for more." It generally starts when the client company asks for more service. The vendor then responds by asking for more money to cover the cost of the service. This cycle continues until contract overruns grow significant enough for management to notice. Suddenly, the outsourcing deal doesn't look so cost effective after all.

In some cases, the nickel-and-dime syndrome starts almost before the ink is dry on the outsourcing agreement. If there was intense competition in bidding on the contract, the vendor may have purposely underbid the true cost of providing the service. Once the contract is in hand, the vendor starts asking for more money every time the client makes a service request.

One Texas-based energy company that outsourced its entire IS operation suffered greatly from the nickel-and-dime syndrome. It got so bad that employees who needed the services stopped asking the designated vendor for them. Instead, employees found back-door ways to hire other contractors to provide the service, negating any savings that the outsourcing contract would have generated.

Very closely related to the nickel-and-dime syndrome is the problem of escalating costs, even though the client doesn't necessarily ask for anything extra. Additional costs may be incurred for equipment upgrades, software licensing, system management, and similar items. These unplanned expenses can quickly drive up the overall cost of services.

If you choose to outsource, you should keep one thing in mind throughout the life of the contract: The vendor is in this deal to make money. If you ask for any services that aren't specifically identified in the contract, the vendor is going to ask you to pay for them. It's that simple.

Contract Termination Problems

It's a fact of life that some outsourcing agreements will end up in divorce court. In anticipation of this, the contract should provide for graceful termination without significant penalties. A few companies found out the hard way that it is very costly to separate from a vendor, for whatever reason.

When Integra Financial Corporation and Equimark Corporation merged in 1992, each bank had its own outsourcing agreement. When the newly merged company decided to discontinue one of the contracts, it wound up having to pay the losing vendor a $4.5 million termination fee.

One way to avoid such a problem is to break the contract up into short renewal periods, say a year or two. Since performance is often cited as a reason for termination, conduct frequent performance reviews to make sure that the vendor is on track with your expectations.

Loss of Control

With outsiders responsible for your day-to-day operations, it's easy to lose control over how things are done. This problem is only amplified if the outsourcing vendor does not share your offices with you. There's truth to the saying, "Out of sight, out of mind."

Keep in mind that you remain responsible for the overall quality of operations, even though the day-to-day responsibility belongs to the outsourcer. Moreover, although outsourcers should be permitted to define the tactics for your IS operations, you must keep control of the long-range strategies.

A good line of communication is essential if you want to retain control. Make sure that the outsourcing agency is serving *your* needs and not its own.

Bureaucracy

Many companies think they are going to eliminate costly and frustrating bureaucracy by outsourcing IS and letting someone else be responsible for operations. We contend that the bureaucracy is still there and is frequently used by the outsourcer as a defense mechanism to protect profits.

The author has firsthand experience in this area. While working for one of the country's largest outsourcing companies, she observed that two sets of management existed: one group from the client side, and one group from the vendor side. The client management team was responsible for directing and overseeing the vendor's activities. The vendor management team arbitrated disputes over the use of resources and oversaw the quality of work performed. This latter team was also responsible for controlling costs. Often, that meant telling the client that he couldn't have the service that he requested, unless, of course, he was willing to pay extra for it. Just when the client thought he could say, "Just go do it!" the vendor management hit him with another bill.

As long as there are conflicting interests, there will always be bureaucracy to protect those interests.

Loss of In-house Expertise

Most organizations that outsource all or part of their IS operations end up letting go of the internal staff that previously did the work. This loss of knowledgeable, skilled employees is a real concern for many companies that outsource. It severely limits the company's ability to bring the work in house again some day.

Even if a majority of the released employees take jobs with the outsourcer, those people are no longer directly under the company's control. Indeed, it's possible now that these skilled workers can be assigned to projects for the company's competitors.

Industry consultant Michael Hammer is concerned about the loss of "technical insight." He feels that internal IS personnel are the only employees who can determine how a company's information resources should be deployed.

Conflicts of Interest

We've already touched on conflicts of interest to a small extent. They arise from the fact that there are now two agendas for IS. Your agenda is to serve your company's business needs—to sell widgets, treat patients, process claims, whatever. The outsourcer's agenda is to make money.

In an ideal world, the outsourcer would be considered your friend, your partner, your advisor. In reality, the outsourcer is all of those things, as long as they don't interfere with making money. If there is a conflict, guess whose interests will be served.

Tips for Successful Outsourcing

Now you know the good and bad of outsourcing. Managed wisely, it can be a very constructive way to hold the line on IS costs and still receive excellent services. If we've piqued your interest enough to make you consider outsourcing for some or all of your IS operations, here are a few tips to success.

1. Look for an outsourcer with proven expertise in your industry and a thorough grasp of your business processes.
2. Be sure to leave yourself some options for getting out of the contract without much difficulty. You may find that the partnership just isn't

working out, and a divorce over "irreconcilable differences" may become necessary.

3. Be sure to retain the right to all source code for applications developed for your company. If you don't, you may find it difficult to maintain the applications later.
4. Stress that all software used by the outsourcer should be unmodified. Then, if you need to transfer operations to another vendor, you should have minimal problems in getting the software to work for you.
5. Work out a very detailed contract before turning over your operations. Good legal counsel is absolutely essential and can save you a lot of aggravation.
6. Clearly define all performance parameters and the penalties for not meeting them.
7. Establish a well-defined reporting structure so that the vendor can keep you informed on issues and progress. The vendor should allow you to establish direction and control of strategic work.

If you and your vendor jointly develop a very detailed contract, you should have few problems in establishing and maintaining a fruitful relationship.

Management Options

In this chapter, we looked at outsourcing as a means to save money, gain expertise, and focus more on business operations. We've seen the pros and cons of this management tactic and reviewed a few tips for success. In the next chapter, we'll look at a few other methods for achieving the same or similar goals. Insourcing, contracting, and consolidation all represent better ways to manage the IS resource.

Chapter 11

Insourcing, Contracting, and Consolidation

Insourcing, contracting, and consolidation are three additional ways to cope with the belt-tightening taking place in many IS shops. They are all methods for managing IS resources.

In Chapter 10, we discussed outsourcing—farming out IS work to a third-party provider. Insourcing is just the opposite—keeping or bringing the work in house and controlling all the resources. While this is not the trend today, there are some instances where insourcing is preferred.

Contracting refers to bringing in "hired guns" to fill specific IS needs. In contrast to outsourcing, contracting allows you to retain full responsibility for the tasks at hand—you just use outside labor to do the work. Contracting gives you tremendous flexibility in resource leveling. That is, you can hire just the right number of people with the right skills at the right time for a project. When those resources are no longer needed, you can send them packing until the next job comes up.

And then there's consolidation. After years of gradual growth in the IS department, many companies find that they have too many people and too much equipment spread out over too many areas. There usually is a lot of duplication of effort, along with equipment, skills, and other resources. Consolidation is a way to rein it all back in.

Many organizations find themselves turning to one or more of these tactics to cope with the changes and challenges within IS. In this chapter, we'll discuss when it is appropriate to turn to each option, and when you should consider other alternatives.

Insourcing

Outsourcing is like sending your child off to boarding school. You use someone else's resources (the school's) to complete your task (teaching your child). Insourcing, on the other hand, is like home schooling. You use your own resources to accomplish the goal. Companies that insource are looking for better ways to use and leverage internal business and technology resources. Unlike outsourcing, the companies themselves retain control over all activities and resources.

Perhaps the best way to explain insourcing is to give an example. Suppose we have an organization that is installing a lot of desktop workstations. Insourcing means that the company does all of the work itself, from burn-in of the hardware to installation on the users' desks. Activities probably include taking delivery of a "raw" PC, burning in the hardware components, loading the operating system, installing additional memory and processor chips, installing network cards and other internal devices, installing application software, configuring all hardware and software for the user's needs, delivering and setting up the PC on the user's desk, running network cabling, and making repairs to the system as needed.

The trend today is for most companies to move *away* from insourcing. Managers find that insourcing IS activities requires too many resources, particularly people with expensive skills. Only very large organizations that can achieve economies of scale attempt comprehensive insourcing. The large Houston-based engineering and construction firm Brown & Root has reportedly benefited from insourcing. Because the company was installing thousands of PCs at its many sites worldwide, it found that using internal resources to set up the PCs was more efficient than using outside vendors.

Selective insourcing is always an option. To continue with our example, a company may find that it's more efficient to let a vendor do the hardware setup, while company employees do the software installation and setup.

There are a few instances when insourcing is a company's best management tactic. First of all, insourcing makes the most sense for very strategic activities, such as application development for a mission-critical project.

Second, insourcing is good if the company can realize significant savings from doing the work in house. Third, you would want to insource if no outside service provider could do the necessary work in a timely manner.

The biggest argument against insourcing is the high investment in resources: people, time, equipment and tools, space, and so on. When the cost of these components is added up, insourcing may not be a very attractive prospect.

Contracting

Contracting has been around since the days of ancient mercenaries. It is the practice of hiring a worker with a particular skill on a temporary basis. Contractors are the "Kelly Girls" of the IS world: They go where they are needed, when they are needed, for as long as they are needed. We are seeing a trend toward increasing reliance on contract IS employees.

There are any number of reasons why a company would use contract employees. Frequently, the company has a project deadline that can't be met by using only internal resources. Contractors can be brought in to fill the resource gap. Or perhaps the company needs a particular skill or certain expertise that no in-house employee has. Government agencies, and some private companies as well, use contractors to get around hiring freezes and other employment regulations.

Since contractors don't normally receive the same benefits as full-time employees, companies can save on total compensation. In fact, it is becoming quite common for companies to lay off employees and then hire them back as contractors. The companies get the benefit of trained employees without having to pay for medical insurance, stock plans, and the like.

Contracting is also great for resource leveling. Contract employees can be hired for as long or short a period as necessary, though usually for less than a year. (For periods longer than a year, it may be worthwhile to hire a person permanently.)

Hiring an employee on a contract basis is also a good way to "test him out." If he proves to be a valuable worker, you can hire him full time. If he doesn't work out, you can let him go and cut your losses quite easily.

Of course, there are drawbacks to using contract help. For one thing, a contractor may have little or no loyalty to the company that hired him. He is working strictly for the money and may choose to leave at an inconvenient time.

Of greater concern, particularly in the application development area, is that when the contractor leaves, the knowledge of the project also leaves. If an application is not well documented, it can be very difficult to support once the developer is gone.

If you decide to use contract help, where can you find the appropriate people? You have a number of resources at your disposal. First of all, there are contract houses (sometimes called "body shops") that have people for hire on staff. While you are most likely to find the particular skills you need through a contract house, you will probably pay more for these workers than for others, due to high administrative overhead.

Former employees also make good prospects for contracting. The biggest benefit here is that they know something about your company already, so they'll probably spend less time getting up to speed for the project.

You can also advertise in the local newspaper for contract help or work through headhunters that handle IS job placement. For small projects, you can spread the word that you are looking for someone through professional organizations like the Data Processing Management Association.

Consolidation

This past decade has definitely seen an increasing trend toward data center consolidations. For most companies, the move is an effort to reduce costs and regain control of computing resources. The 1970s and 1980s had seen a slant toward distributed data processing. At the time, it was more efficient to have processing capabilities physically located with the end user. With the advent of high-speed data communications, it is no longer important to place computing resources where the consumers are. Companies are beginning to consolidate far-flung resources and reduce duplication of staff, equipment, and facilities. In recent years, consolidations have eliminated one of every ten data centers.

Data centers are not the only facilities that are ripe for consolidation. Royal Bank of Canada recently consolidated its help desks, going from more than 90 down to just three. Prior to the consolidation, a separate help desk existed for just about every application that the bank had. Help desk manager Fatima Leathwood reorganized support and distributed the help desk tasks geographically, rather than according to application. The move resulted in greatly reduced costs, increased efficiency, and dramatically improved customer satisfaction.

Network services can be consolidated as well. In December 1992, the Federal Reserve Bank (the Fed) announced plans to create a national communications backbone network that will allow the organization to reduce the costs of network services. The new network will allow the Fed to merge data processing centers in the 12 regional Federal Reserve Banks into three data centers.

While cost reduction is most often the motivation behind consolidations, cost avoidance is frequently a goal as well. As computing environments grow more complex, companies are likely to require larger data center staffs. Consolidating the resources and reducing the diversity of systems can help stave off IS budget increases.

Automation tools are critical to the success of these new "super centers." These centers often have tight processing schedules and are in operation 24 hours a day, seven days a week. Automated systems regarded as critical to managing data centers are tape management systems, job scheduling and balancing systems, and console automation tools. The tools enable staff reductions, allow operations to be managed remotely, and make data center operations more cost efficient.

While automation technology is a necessity for a well-run shop, it is also good for a company's image. Bankers Trust Company is using its data center consolidation to show current and potential customers that it is committed to world-class technology. The bank uses floor-to-ceiling glass walls to exhibit the company's new computer command center.

Consolidation efforts are often politically charged, particularly for government agencies. In 1991, the Air Force began choosing sites for consolidated data centers. The plans called for several dozen existing sites to be whittled down to a half dozen or less. Groups that stood to be affected by the consolidation lobbied Congress to spare their facilities. One official involved in the project claimed that, although four data centers were originally planned, a fifth or even a sixth might be added because "somebody's bacon needs to be saved."

Politics play a role in corporate consolidations as well. Which sites will be eliminated? What staff members will be spared? Who will be left to manage the remaining resources? These and other questions are often settled on the basis of the strengths of the personalities involved.

Resource Management Options

These days, the IS manager has a lot of options for managing his resources intelligently and holding the line on the IS budget. Insourcing, though not very common today, remains a viable alternative for the IS department where a "do-it-yourself" attitude prevails. Contracting is a practical option when special skills are needed on a temporary basis. Finally, consolidation helps an IS manager combine and centralize resources and facilities.

In the next chapter, we'll look at downsizing and rightsizing, which both deal with an organization's application platform. Both strategies present many opportunities for saving money and improving application performance.

Chapter 12

Downsizing and Rightsizing

If you start hearing corporate management throw around the buzzword *downsizing*, listen closely to how it is used. It doesn't necessarily mean you'll be getting a pink slip soon. The term downsizing has two different meanings in the business world today. In a broad business sense, downsizing refers to making an organization smaller, usually through employee layoffs and consolidations. In the realm of information technology, downsizing refers to moving all or part of a computer application to a smaller computer system or a network of desktop computers. This book addresses the latter of the two meanings.

The words *downsizing* and *rightsizing* are often used interchangeably, although rightsizing can have a shade of meaning of its own. Rightsizing properly designates the process of placing a computer application on the hardware platform that makes the most sense for it. This platform can be larger or smaller than the original host platform. For instance, a database that was developed for a standalone PC can be rightsized (or *upsized*) onto a network platform when more people than one need access to the application. For purposes of simplicity, here we use rightsizing and downsizing to mean the same thing: moving applications from "large" boxes to "smaller" boxes. More specifically, our choice for a smaller platform is a network of microcomputers and workstations.

Downsizing is an important trend in information systems today, although it is not a new concept. In the late 1970s and throughout the 1980s, companies downsized from mainframe computers to midrange or minicomputers. The process today more commonly involves moving applications from the mainframe or minicomputer down to workgroup-oriented computing platforms such as superservers and networked desktop computers.

In this chapter, we look at why downsizing is so important, and at what has enabled the trend. We'll discuss some of the benefits to downsizing and look at a few case studies of companies that have gone through the process. Finally, we'll examine the strategies for moving applications, explore some of the obstacles to downsizing, and share our steps for success.

Why Downsize?

According to a recent survey, up to 80 percent of Fortune 1000 companies are actively engaged in downsizing projects or plan to get started soon. Why is everyone so eager to jump on the bandwagon? The most common reason cited is to save money. Many companies have come to realize that mainframe-based computing is simply too expensive, and that there are less costly computing alternatives.

A second commonly cited reason for downsizing is to increase a company's ability to adapt information technology to business needs quickly—in other words, to increase flexibility. The business environment is changing so rapidly these days that mainframe-based information systems simply cannot keep up with the pace of change. Traditional large-scale applications take years to design and to develop. In today's business world, these applications may be obsolete before they are ever implemented. By contrast, applications developed for network computing using such new tools as Visual Basic and C++ can be developed relatively quickly. Moreover, network platforms offer "plug and play" flexibility, whereby additional capabilities can be added quite easily.

Some companies have turned to downsizing applications in order to give control of the corporate data to the business people who need it. In the mainframe environment, IS controls the data and closely guards access. Only those people can freely access the data who are technically adept at using the arcane languages and commands of the mainframe environment. In the network environment, mere mortals can access the data by using common, familiar tools such as spreadsheets and query-by-example products. With easy access to data, the business unit manager gains the power to make business decisions based on factual and accurate information.

After years of investing millions of dollars on desktop computing facilities, companies are looking for the productivity payback. By downsizing to network-based computing, these companies can realize their potential for higher productivity with "friendlier" tools and techniques. For example, new graphical user interfaces can make applications easier to learn and use. Even complex data entry and query screens can be made more appealing with features that are unavailable with traditional mainframe products: cascading menus, context-sensitive help, data pick lists, and more. PCs and workstations can provide almost instantaneous response time, compared to mainframes with terminal connections that can take several seconds or more to respond to each request. As a result, the network environment is conducive to putting more applications on a computer. The computer becomes an extension of the individual worker, providing him with fast access to powerful tools that turn data into information. It is not simply the warehouse for vast, detailed data.

Another reason that organizations decide to downsize is to "get open." Frustrated with the boundaries that proprietary large systems enforce, many companies are turning to open-architecture computer systems. In a broad sense, open systems are those that have support from a lot of manufacturers. This means that no one vendor is controlling standards or the future of the technology. And, technology can be recycled and reused. The servers of today might become the desktop computers of tomorrow. Companies are not forced to put all of their technological eggs in one basket, or to declare nearly new equipment obsolete.

Some companies opt to downsize applications in order to stave off an expensive upgrade to the mainframe. By removing select applications from the mainframe, they make more capacity available for the systems that simply must be run on it. Good candidates for rehosting are non-critical systems such as those for human resources and accounting. There are plenty of off-the-shelf network-based packages that are suitable for handling these needs, making the transition to the new platform easy. In the mean time, mainframe facilities are freed, helping to delay or avoid an upgrade in power and capacity.

The Enablers for Downsizing

Application downsizing has become a hot trend only in the last few years. Up until then, it just wasn't practical to rehost mission-critical applications on a network. Some factors have enabled the move to these smaller computing platforms.

Network Maturity

First of all, network technology finally offers the power and capacity that are needed by major business systems. Workstation and superserver processors can easily match the processing power of mainframes and minicomputers. Measured in MIPS (millions of instructions per second), servers based on Intel 80486 or Pentium chips can process as much information as the powerful IBM 3090 mainframe. In terms of transactions per second (TPS), computers based on RISC (reduced instruction set computing), CISC (complex instruction set computing), SPARC (scalable processor architecture), or parallel processing technology can easily meet or exceed the capabilities of mainframes.

NCR Senior Vice President and chief scientist Philip M. Neches claims that microprocessors have increased in power at an annual compounded rate of between 35 and 40 percent. In each three-and-a-half-year generation, microprocessor performance quadruples, while mainframe performance doubles. With every generation, microprocessors take two steps toward mainframes in terms of absolute performance. Says Neches, "A decade ago, the gap between mainframes and microcomputers was very large; today it is almost gone."

In terms of capacity, network storage devices can handle multiple gigabytes of data. Database products are designed to process virtually billions of records. And, network bandwidth and throughput can satisfy most applications, except perhaps where video information is concerned, but even that is changing.

Network operating systems have evolved from small workgroup systems into enterprise systems. Products like Banyan VINES and Novell NetWare 4.0 can accommodate hundreds of users on a single network, offering users ready access to resources and network administration.

Network hardware has achieved very high levels of reliability. Server products in particular incorporate fault tolerance, disk mirroring and duplexing, automatic error detection and notification, and many other features that keep the system functioning 24 hours a day. These products are as reliable as the mainframe, perhaps even more reliable. Moreover, network products are often easier and cheaper to fix when something does go wrong.

Other tools and products needed to support mission-critical applications on a network have matured as well. Client/server front ends and back ends are robust enough and stable enough to host truly critical applications. User interfaces and application development products are feature-rich and powerful.

All in all, network technology has matured to a level where it has the power, capacity, and reliability to accommodate enterprise applications. This fact has been a great incentive for IS managers to finally move their applications off the mainframe. They can now feel comfortable about the stability of the new platform.

Cost of Ownership

Another enabling factor in the downsizing movement is the overall cost of ownership of networks versus that of mainframes. One astounding comparison is the cost of MIPS. For an IBM 3090 mainframe, the cost to process one MIPS is between $94,000 and $105,000. The same measure of processing power costs around $200 to $500 on a 80486-based PC.

For the purist who argues that using MIPS to compare PCs to mainframes is not valid, Larry Ellison, President of Oracle Corporation, suggests comparing the cost of TPS on the various platforms. Ellison believes that this is a valid measurement, since the same program can be run and benchmarked on different systems. When compared, the cost per TPS is at least 20 times higher on a traditional mainframe than on a microprocessor-based system.

The cost of microprocessors continues to fall, helping to bring down the overall cost of networks. For example, *Computerworld Magazine* reports that the price for Intel's 33-MHz 486DX chip fell 37 percent from December 1990 to December 1991. By comparison, the Intel 386 chip had been on the market for five years before it showed a similar one-year decline in price. This comparison indicates that the trend toward lower-cost, higher-powered computing continues to accelerate.

Beyond the cost of components, mainframe computers have high requirements for operation and maintenance. Most mainframes require a tightly controlled environment with low humidity, a specific room temperature, cooling facilities, and tremendous power. They also occupy a large amount of expensive space. By comparison, a network server and associated workstations can be located almost anywhere, under much more extreme conditions than the mainframe. As you can imagine, the mainframe's operational requirements add significantly to the cost of ownership.

Support personnel are also expensive, and a mainframe environment tends to require more people than does a network environment. Since salaries are the top expense in most IS departments, reducing the number of support people can translate into higher cost savings.

Software license fees for the mainframe routinely run into many thousands of dollars. For instance, a mainframe-based relational database management product can cost up to $300,000, with annual maintenance fees running as high as 15 percent. A comparable database system for a network costs about one tenth that amount.

Since most companies downsize in order to save money, the overall cost of ownership of a computer system is very important. It is not unusual to hear of companies that are saving millions of dollars each year by moving applications from mainframes to network platforms.

People Are Ready for Desktop Computing

People, also, are enabling the downsizing movement. Not only is the network platform finally ready for "prime time," but people are ready for networks to happen. Organizations are changing from hierarchical structures to workgroup-oriented structures. These workgroups are accustomed to collaborating through the use of networked computers—with e-mail, with work-flow software, with databases.

Our society is becoming a more computer-literate society. Most white-collar positions now require use of a computer, and companies are moving toward a one-to-one ratio of PCs to knowledge workers. Thus, employees are not willing to wait for someone in IS to give them access to the information they need. In terms of computing, employees are ready to do more for themselves.

Who Is Downsizing?

The 1992 *Computerworld* study of the Premier 100 (the top 100 corporate users of information technology) shows that the main challenges for IS executives are "reducing costs, managing more work with the same resources, educating business users, meeting strategic corporate objectives, and downsizing." Interestingly, successful downsizing can help meet the other challenges.

So, what companies are downsizing? The question might be more easily answered if we ask, "What companies aren't downsizing?" Most of the Fortune companies (and many not on the Fortune list) are downsizing at least some applications. And the trend isn't just for corporations. Government and educational institutions are also jumping on the bandwagon. We could cite hundreds of case studies, but we'll keep to just a few.

Downsizing Case Study: A Financial Services Firm

In 1987, an information overload led Financial Guaranty Insurance Company (FGIC) to downsize. Andrei Chivvis, First Vice President of Systems Services, says that the problem had reached crisis proportions. The insurance and investment banking company, founded in 1983, had grown rapidly. Consequently, FGIC outgrew its IBM 4381 mainframe's capabilities in only three short years.

The information systems problem was beginning to threaten the development of the business. The long development cycles for mainframe applications couldn't keep up with the company's rapid changes. FGIC management acknowledged that there were two alternatives: spend more money to expand and improve the existing mainframe-based systems, or convert to another computer system.

FGIC made the bold decision to move all of its systems to a decentralized PC-based network. Chivvis felt that three major factors helped him to feel comfortable with this decision. First of all, the Intel 80386 architecture had become available and provided adequate throughput and growth potential. Second, PC-based network operating systems had reached a stable performance level. Finally, versatile application development tools made database development easy and flexible. Chivvis claims that the many innovations in the microcomputer market helped him to sell the LAN idea to upper management.

From the outset, FGIC faced an important deadline. Government and client reporting requirements had to be met by the end of 1988. This gave Chivvis and his crew just 18 months to completely rewrite all applications from scratch. FGIC found that it had to undergo a fundamental change in business processes as well. For instance, records on the old computer system were frequently out of date because of a sluggish data entry process. The new system would permit financial analysts to do their own data entry as they created a business deal. This process kept the data in the system from being out of date and provided management with better control of the data.

As FGIC selected the network hardware and software, a decision was made to go with best-of-breed components from different manufacturers. Chivvis's group chose technology that was embraced by several manufacturers in order to achieve flexibility through multiple options. Chivvis likes the idea of being able to change out components quickly in order to adapt to changing needs.

When it came time to implement the new system, Chivvis advised his team "not to throw out everything learned from the mainframe." Despite their flexibility and open systems, networks must still provide security procedures, have adequate backup, and allow for disaster recovery. In these areas, Chivvis says, his network is "better than the mainframe." Because LAN components are relatively inexpensive, he claims that "it doesn't cost much to have equipment off site and be ready to go" in the event of a disaster.

The network has created quite a cost savings for FGIC, as well. The cost of the mainframe, with fewer capabilities, was about $6 million per year. The network, with more capabilities, is costing about $2 million per year to operate and support. The biggest benefit, however, says Chivvis, is that "we can now respond to any new business requirement almost as soon as our business people can formulate what they want. The generation gap between needing an application and developing it is gone."

Downsizing Case Study: A Banking Institution

A PC-based network permits the Richmond Savings Credit Union in Vancouver, British Columbia to take an innovative approach to providing a wide selection of banking services. Richmond Savings offers "relationship banking" to its 48,000 customers. It has bounced back from flat performance to a rapid growth position, with more than $900 million in assets in some 200,000 accounts. The key element in the credit union's comeback is the client/server database technology that is used to organize all customer activity.

Richmond Savings implemented its original network in 1987, at a time when LANs were not considered adequate to mission-critical applications. A 10 MHz, AT-compatible PC served as the central database server for all vital customer information. (The server has since been upgraded to a 486-based PC, and the original 286 PC has been recycled as a workstation.) Each of the credit union's eight branch offices has its own local server to process requests by the local workstations. (More than 300 workstations are currently on the network.) However, all database information is still stored at the central PC.

The functionally rich system shares online connections to external systems such as automated banking machines, clearing systems, and other databases. It also supports a parallel office automation system. The system processes an average of 65,000 online transactions daily and offers average response times under three seconds.

Allan Lacroix, Vice President of Technology, says that the old mainframe system (which was completely replaced by the PC network) was not as robust. The 3GL development environment made applications difficult to change. The PC network has simplified application development and has allowed Richmond Savings to respond rapidly to a changing business environment.

Richmond Savings embraces the concept of using business analysts to design and develop the applications. Says Lacroix, "We find it is better to take someone from the business side and teach him how to program than to take someone from the programming side and teach him the business." Many of Richmond Savings' original LAN applications were developed by an experienced teller using the network database management system and its associated tools.

The credit union definitely views this network as its competitive advantage. Lacroix says that the institution's culture changed significantly after the network was implemented. For instance, when a customer comes to the credit union to request a loan, there is no "paper" application. The loan manager (now known as a "personal banker") uses the desktop workstation as a part of the interview process. The loan request is filled out entirely on line. The next time the customer comes in for a transaction, his portfolio information is readily available.

Downsizing Strategies

There are several strategies for getting started with a downsizing project: porting the application as is, rewriting the application, purchasing a packaged application, and reengineering a process and its associated information systems. The optimal strategy depends upon many factors, including time (the urgency of downsizing), cost, the availability of packaged applications, the knowledge and skills of in-house programmers, and more. In general, however, the payback increases as applications are redesigned and rewritten.

Port in Place

Perhaps one of the quickest ways to downsize an application from the mainframe to a network platform is to simply port, or rehost, the application as is. This strategy requires the fewest changes to an information system. In this instance, the company changes the hardware platform, but not the software.

Products such as Oracle and Focus that span a range of platforms help to make porting in place a quick solution for getting off the mainframe. Moving the data from the mainframe to the LAN most likely requires little or no conversion. A major consideration in selecting this strategy must be the level of flexibility that is desired in the "new" system. For instance, porting a COBOL application from the mainframe to a COBOL environment on the LAN may save money, but it doesn't add a great deal of value to the application. The recompiled COBOL code does not exploit the advantages of graphical user interfaces and other productivity aids that are popular on the desktop. Moreover, simple rehosting does not incorporate client/server computing, where most of the advantages of downsizing are realized.

Purchase a Package

A second strategy for moving an application is to purchase an off-the-shelf package to meet the company's needs. There are many packages available for common applications such as accounting, payroll, and human resource management. Most can be adapted to the company's specific needs with little or no modification. As for highly specialized or vertical applications, packages just don't yet exist for the network platform.

Ready-to-use packages are often developed using the most current tools and techniques available. They frequently employ a graphical interface, which end users will generally find appealing and easy to learn and use. Off-the-shelf software can be purchased for as little as a few thousand dollars. The time and cost of moving or converting the data, however, could be significant.

Companies with little experience with downsizing, and particularly with application development on the new platform, should consider purchasing canned applications. Downsizing involves many changes, but, if off-the-shelf software is available, learning new development techniques doesn't have to be one of them. Moreover, the effort of building new general systems is hard to justify when they are available as near-commodity items.

Rewrite the Application

A third option, which appears to be the most common among companies that have downsized to networks, is to rewrite the entire application. With today's changing business climate, the life cycle for a typical information system is only a few years. Thus, it is likely that the mainframe-based application is outdated and in need of enhancement anyway. Rewriting the application gives the company an opportunity to add new features and functionality that may not have been required or even possible before. While a complete rewrite of the system is time-consuming and costly, it is often the most effective option for meeting the users' needs.

Rewriting the application also gives a company the opportunity to use the latest development languages, tools, and techniques. Chances are, the legacy systems were developed using COBOL or some 4th generation language, both of which require significant computing overhead and offer little opportunity for reusing precious code. By comparison, applications developed using object-oriented programming techniques are much more efficient. What's more, object libraries can be used to store reusable code, reducing development time for ensuing applications.

Reengineer the Process

A fourth strategy for downsizing is to completely reengineer the business process before developing the information systems to support it. This is more than a simple "rewrite" of the application; rather, it is a redesign of how the work is accomplished. Chapter 9 addresses the whole area of reengineering.

Once the new business process has been mapped out, the company can select how best to implement the supporting computer applications. The company can choose off-the-shelf packages or develop home-grown systems. This downsizing strategy usually offers the greatest benefits in terms of cost savings, increase in flexibility, and leverage of information technology.

Common Pitfalls of Downsizing

Companies that have gone through downsizing exercises report that the sailing isn't always smooth. There are some pitfalls to watch for, and fortunately, you can benefit from everyone else's experiences. We have collected some of the common complaints, listed below.

- *The network requires more administration and control than expected.* Moving from a highly centralized mainframe system to a distributed network environment places more administrative demands on more people. Most companies with networks employ departmental administrators, making coordination among the administrators essential for smooth operation. New enterprise-network products like NetWare 4.0 should help to alleviate this complaint.

- *Integration in a multi-vendor environment can be frustrating.* Sad to say, the days of buying a complex computing system from one vendor are all but gone. In the era of open network computing, we will always be dealing with integration problems in a multi-vendor environment. Troubleshooting, in particular, can be a real challenge as vendors point the finger of blame at each other's products.
- *Good network management tools are hard to come by.* Network management is always a challenge, especially when the network incorporates disparate systems. Effective management tools are slowly coming to market in a piecemeal fashion.
- *Network security is not as easy to maintain as security on the mainframe.* This is not to say that network security measures don't exist; they do, and there are plenty of them. With processing distributed across a vast network, however, there are more opportunities for security breaches.
- *Some technologies are too close to the "bleeding edge" to implement just yet.* There's an old saying in the IS industry: Never buy version 1.0 of anything. Instead, wait for the next release so the product's problems can be worked out. Downsizing pioneers took chances with all kinds of new technology. While companies occasionally have to take risks to obtain early payoffs, they should minimize their exposure to untested technology.
- *There is a dearth of qualified people with the necessary skills and knowledge.* Companies that are adopting new technologies with their downsizing projects often find it difficult to find qualified help, particularly in the area of client/server applications.

Steps for Successful Downsizing

After years of observing the downsizing market, Currid & Company has formulated its Ten-Step Plan for Successful Downsizing. We share those tips with you now.

1. Establish a business objective for downsizing.
2. Secure top management support, especially from the CIO.
3. Handpick the implementation team.
4. Build a strong network platform.

5. Assess the desktop environment.
6. Select a strong client/server back-end database management system.
7. Find a good client/server front end.
8. Choose a "downsizable" pilot project.
9. Outsource the mainframe once your major applications have been offloaded.
10. Reorganize IS to support the new computing platform.

Observations on Downsizing

While each organization's downsizing experience is unique, there are some common observations.

- Downsizing often takes place as a response to financial problems, new personnel changes (at the executive levels or head of IS level), or new computing procedures sought by the line of business managers.
- Downsizing is not an overnight process. For those companies that ultimately replaced their mainframes entirely, the process took from two to five years.
- Downsizing frequently requires significant IS restructuring. Most IS departments shrink, although some have grown to service new information needs.
- Downsizing frequently places more power in the hands of end users. Whether it be data control or report writing responsibilities, many companies find end users begin to take on new tasks. This forces computer literacy upon some corporate workers.

In this chapter, we looked at the common reasons for downsizing and the factors that have enabled the trend in recent years. We explored the strategies, the pitfalls, and the steps for success. In upcoming chapters, we'll take a closer look at the technologies behind downsizing, including LANs, WANs, and client/server computing.

Chapter 13

LANs and WANs

Thomas C. Haliburton once said, "Death and taxes are inevitable." Well, we should probably remind Mr. Haliburton about change. In the business world, you need to adapt and overcome the obstacles and new situations that the chaotic world of free enterprise throws your way, because there is no avoiding change. If you or your organization cannot acclimate to a new economic environment, then you will be left behind.

We have already taken a good look at the trends molding business ventures in the newly globalized and highly competitive world economy. We have considered the need for changing the way companies are organized and operated. We have even discussed the necessity for accurate and timely information that will help your organization maintain an edge in today's competitive marketplace.

But, the question still remains, how are you going to provide a conduit that will connect the people of your organization and allow information to flow to and from the decision makers who need it? Also, what tools will you provide to help increase the productivity of your workforce?

The answer is simple: technology. However, the technology or the tools that we use to do work must also be adaptable, since they too are subject to change. The adaptability of desktop computers, especially if they are networked, has made them one of the more important factors in modern corporate life.

In this chapter, we will discuss what LANs and WANs really are and some of the history leading up to the explosion of PC networking. We firmly believe that this technology is fundamental to an effective organization in the '90s. We will also discuss the features that make LANs and WANs a valuable business tool.

What Are LANs and WANs?

Local Area Networks (LANs) and *Wide Area Networks (WANs)* can be an immense advantage to your business. They can provide you with the means for gathering, processing, and distributing the data that your workforce needs. They can also be the means by which you and your coworkers can improve your efficiency. But what are LANs and WANs really?

A LAN or Local Area Network is really nothing more that a group of computers attached together in such a way that they can exchange information and share resources. A LAN is not limited to PCs or desktop computers but can include minis and mainframes as well.

A WAN or Wide Area Network is basically a network of LANs that may or may not be located in the same area. In fact, a WAN can often connect machines located on totally different continents. This flexibility of both machines and locations is one of the more important features a LAN or WAN can provide you.

Business Computing: Then and Now

As was mentioned earlier, the '60s and '70s were the ages of the mainframe and the minicomputer, respectively; they were shortly followed by the desktop and network computer eras of the '80s and '90s. Networking PCs is definitely the current trend. According to International Data Corporation, 43 percent of the PCs used in business settings were networked in 1991. This number jumped dramatically in 1992 to 60 percent. It is now estimated that 90 percent of all business computers will be networked by 1995. Why is this the case? What is so special about networks of PCs that has helped them replace mainframes and minis as the technology of choice?

In order to understand more about how LANs and WANs are useful, we need first to discuss a little of their evolution.

Mainframe Computing and Minicomputing

Research of any sort is ultimately going to pay off. The technology developed for one particular research project could have an infinite number of uses not envisioned in the original project's scope. So it was with computers. The earliest computers and the accompanying technology were developed to advance atomic research and, more specifically, the nuclear weapons program, since this new technology so greatly aided in carrying out the math and physics needed to develop nuclear bomb models.

Like so many other new technologies, the early mainframes needed to have their rough edges smoothed before they could be used for more mundane purposes. The early computers were monstrosities used almost exclusively by the scientific community and the government, simply because they were so difficult to use and expensive to purchase and maintain. In fact, ENIAC, one of the more celebrated attempts at a computer, was extremely difficult to use and maintain. ENIAC inoperative more often than not. Special teams of technicians were needed to constantly search for and replace bad vacuum tubes. Others had to physically rewire ENIAC to perform each new function that was devised for it. Also, the heat ENIAC produced kept the temperature of the building it was in at a comfortable 120°F. The '60s were to arrive before computer technology had advanced far enough for mainframes to be used in the corporate scene.

The mainframes were more reliable but still had many of the problems illustrated above. They were still very expensive to purchase and maintain, requiring the special environments or glass houses mentioned earlier to counteract the heat output of the mainframe. Mainframes no longer needed to be "hard coded" like ENIAC did. Instead, they could be programmed using "software." However, this still required specialized staff to write programs to manipulate the data. The applications they wrote were vertical, expensive, and inflexible, as well as having long development cycles.

Despite all these shortcomings, mainframes did provide corporations with computing systems that were centralized and characterized by high levels of security, fault tolerance, and manageability. This centralized computing model fit in well with the hierarchical corporate structure prevalent at this time and managed to create a niche for itself during the '60s. Minis were an attempt to broaden the appeal of mainframe computing by making a smaller, more affordable mainframe.

Desktop Computing

In the late '70s and early '80s, we began to see a new direction in the computer world with the introduction of the *IBM Personal Computer,* or the

PC as it became commonly called. The IBM PC and the inevitable copies or *clones* of its basic architecture and design revolutionized the business computing environment. Instead of each user sharing the processing power of an expensive mainframe, each PC user had a microprocessor all to himself or herself. Even more important, the operating system, DOS, was developed by a third party (Microsoft), making it easier for other interested third parties to develop software. Users didn't have to rely on IBM to develop all the applications for the PC. Anybody with an idea for a useful application could create and market the software. This intense competition in both hardware and software innovation produced a cheap, easily usable tool that seemed tailor-made for the business world.

There were drawbacks, however, to the office full of PCs. For one thing, resources could not be easily shared. If you needed a specific printer, you had to move the file you wished to print to the PC the printer was attached to. The same problem existed with information. In sharing files, special care had to be taken so that changes made by one person were not overwritten by changes made by another. Data security was also a problem, with backups of important data being at the mercy of the individual user.

Despite these drawbacks, PC computing began to gain in popularity for many reasons. PCs were cheap to purchase, and almost anyone could learn to operate one. PC software was geared to the end user, allowing the workforce to accomplish more by itself without having to rely on the mainframe "gurus" to generate the information needed. PCs provided small business people with the power of computers previously denied them by the heavy expenses of mainframe and minicomputing.

Why Network Computing?

The next obvious step for business computing was the fusion of the stability and security of the mainframe environment with the flexibility, ease of use, and the economy of the PC. This, of course, was accomplished with PC networking. Today, a LAN or a WAN can provide you with:

- Increased access to valuable information
- End-user empowerment
- The ability to share valuable resources
- Structural flexibility
- Increased data security

Information Access

In today's business jungle, the quicker information is obtained, processed, and delivered to those who need it, the better. A network can serve as the vehicle by which this information is quickly gathered and passed on to those who need it through programs and services such as e-mail, group schedulers, fax service, databases, and many others.

For instance, an important client wants to set up a meeting with you and some other important decision makers of your firm. Instead of wasting time calling everyone and coordinating everyone's schedule, you could consult a group schedule program that just happens to be on the network, and that would tell you the best time for everyone to meet. You could even reserve a room through the program. An e-mail message could follow to advise participants of the schedule change.

Another example could be a new procedure developed by the engineering department that helps customers to use your company's line of low-cost widgets more effectively. This information could be passed to the customer support department for distribution. Once again, an e-mail message could pass along the information or the location of the information file to the customer support representative in charge of the low-cost widgets. The customer support representative could then access the sales department's database, search for and compile a list of customers who own the product in question, and merge the appropriate customer information into a form letter explaining the new procedure. These letters could then either be printed and mailed to the customer or be faxed directly to the customer without having to be printed. You can begin to see how quickly information can be received, accessed, reacted to, or distributed using a network.

End-User Empowerment

You should begin to realize a few additional important points from these previous examples. In both examples, a network user received some information and was able to perform a wide variety of tasks quickly, using nothing other than the networked PC sitting on her desk. She did not have to find someone to access the data and manipulate it for her, as she would in a mainframe environment. She accessed the data available on the network through powerful yet easy-to-use PC applications.

From the very beginning, PCs, PC applications, and PC network software were meant for the ordinary user. If a software package is not both powerful and easy to use, it will most likely not gain acceptance in the marketplace. This double emphasis means that nearly anyone can understand how to use PC software to access and manipulate data over a LAN. With a little training, almost anyone can be "empowered" to accomplish much in a fraction of the time previously necessary.

Sharing Valuable Resources

Another important advantage to a LAN or WAN is the ability to share resources. For instance, to attach a printer to every workstation would be extremely expensive. But with a LAN, machines can share a printer through network printing services. A file to be printed is sent to a network print queue to await its turn. Which queue the print job is sent to determines which printer will print the data.

A LAN or WAN can also share resources such as fax services, modems, and optical disks, among other things. As was shown earlier, a LAN or WAN also allows you to share the most valuable resource of all, information.

Flexibility

A LAN or a WAN can provide you with the flexibility to build a corporate computing structure that is tailored to your needs but able to grow as your business dictates. If you have a small office and need to network only a few computers, then a small peer-to-peer network could do the trick for you. If your network is larger than a few computers, a client/server model network may be better. You can have a network of only two workstations all the way up to one thousand.

Networks can provide you with the flexibility to organize your network structure along the lines of your corporate structure. For example, you can provide each department its own LAN. These departmental LANs can themselves be linked by high speed-backbones so that information can be shared between the departments.

If you have offices in different cities, the two different LANs can be connected by modem over normal phone lines or Public Data Network lines to form a WAN. If you need interactive support for your WAN, you can use a high-speed data link, such as a T1 or T3 link, or a geosynchronous satellite link. For example, if you have an office in Los Angeles and an office in Hong Kong, they can transfer data back and forth over ordinary phone lines using modems.

Data Security

Finally, PC networks offer excellent fault tolerance and data security. First, networks normally provide centralized data storage. This makes backing up critical data and applications easy, since all the data are accessible in one spot. If something were to go wrong and data and applications were lost or corrupted, they could be restored easily from regularly kept backups.

Another form of fault tolerance comes directly from the mainframe era. A Redundant Array of Inexpensive Disks (RAID) can serve as the central storage facility. A RAID is a collection of disk drives that dedicates part of its storage capacity to redundant data. If a disk fails, the remaining disks can rebuild its data from data they have stored redundantly.

PC networks also have their own security systems to keep users from accidentally or purposefully destroying or corrupting data. Join these security features with regular backup capabilities and fault-tolerant features like a RAID, and you have little chance of losing important data.

A Business Essential

A PC network has become a necessity in the corporate world because, combining the security and centralized storage of the mainframe environment with the ease of use of the PC, it empowers the end user to be more efficient and productive. In today's fast-paced economy, whoever can react quickest to valuable information will have a huge advantage. LANs and WANs can be a very effective tool for helping your organization collect, analyze, and distribute information to those who need it.

Chapter 14

Electronic Data Interchange

Ever wonder how your favorite department store keeps the shelves properly stocked with neon-lime-green denim jeans with the hole above the right knee? Or how your local grocery store manages to keep just the right amount of frozen concentrated orange juice in the freezer section? In many industries, including retail and grocery, the answer increasingly involves a technology called Electronic Data Interchange (EDI).

Electronic Data Interchange is, simply put, the electronic exchange of business documents such as invoices, purchase orders, inventory inquires, bills of lading and others—virtually any business forms—between suppliers and customers. Although EDI has been around for nearly thirty years, widespread use of the technology is just now being realized by a broad spectrum of industries.

For our purposes, EDI refers to software specifically developed for the purpose of transferring business information that conforms to the Electronic Data Interchange for Administration, Commerce, and Transport (EDIFACT); ANSI X.12; or CCITT specifications for information exchange. One of the key elements of EDI is that it uses these standard formats to grant outsiders and business partners restricted access to information that is useful for conducting business, rather than unrestricted access to a company's data.

In this chapter, we will discuss a little of the history of EDI and review some of the early problems that once prevented its widespread use.

We will describe some of the benefits companies are able to reap through the use of EDI and introduce some of the factors that may affect how it is currently used in your industry. Finally, we will familiarize you with some of the changes in business and government perspectives as well as trends in technology that are likely to affect your use of EDI in the future. The most important message in this chapter is that EDI should no longer be considered a niche technology. We feel that it is on the verge of becoming as pervasive as the business use of facsimile and credit cards.

The Evolution of EDI

EDI began as a niche-oriented application (commonly called a vertical application) in the transport and medical industries, but it has found its way into government, manufacturing, petrochemicals, retail, and other areas as well. As early as 1960, the American Hospital Supply Corporation was implementing purchase order systems that linked them directly to their customers.

The early systems worked well and gave the companies who created and used them a competitive advantage, but they were proprietary in nature. Competitive pressure and lack of standards hampered the widespread use of EDI in its early stages. Even today, because of disagreement about the standards for EDI, the technology is not widely implemented outside specific industries.

On the other hand, within industries where standards are possible or already in place, acceptance of EDI is widespread and growing rapidly. The grocery industry, with the Universal Product Code (UPC), and the publishing industry, with ISBN Library of Congress Numbers, are industries that already have a standard coding method for products and services, and that use EDI extensively. According to the Gartner Group, industry acceptance of EDI is substantial enough for the market to grow between 35 and 45 percent annually.

Dissension among the Ranks

Given the projected acceptance and current popularity of EDI (over 75 vendors currently produce EDI products and services), you might think that everything about EDI was rosy, but the truth is a little more ambiguous. Although there may be many EDI solutions within a particular industry, the unique requirements for each industry make it extremely difficult to share electronic forms among different industries.

Some of the greatest hindrances are related to historical business practices of certain industries dating back hundreds or even thousands of years. According to Torrey Byles of the market research firm Input of Mountain View, CA, historical business practices established thousands of years ago in the maritime industry are simply ignored in the standards, for example.

Another area of distress arises from proprietary information systems that promote "home-grown" Electronic Data Interchange solutions. In the medical industry, patients, providers (doctors), and insurers have long used a wide array of forms and coding conventions when making claims. According to Joseph Brophy, President of Travelers Insurance Co. and Co-Chairman of the Workgroup for Electronic Data Interchange (WEDI), this has led to widespread disparities:

> Insurers have developed sophisticated computer systems to process and track claims. However, virtually every system uses its own standards and formats. Today, there are some 458 claim forms that are essentially identical except for different formats and coding conventions.

The EDI messaging standards are really just formats for storing particular items of information; they are designed to alleviate such problems. The goal is to get everyone using a standard set of forms, which means that the software they use must record data in a standard format. This process, called *mapping,* takes information from a non-standard format and places the information in the relevant locations on a standard form. But the problem is more complex than meets the eye.

There are several different industry-specific forms as well as generic ones in the ANSI X.12 standard, for example. Industry analysts widely agree that, because of differences in software and form content, over 50 percent of the documents shared via EDI have to be re-keyed on the receiving end.

So, where does the rubber meet the road? For now, it is a compromise. Individual industries must rely on conventions within the industry for implementing EDI solutions or establish standards of their own based on national or international EDI standards. Those industries that have already developed a coding scheme for products and services will find it far easier to develop an EDI-based solution than others. In many cases, companies in the industries where the standards have been established desire (and in some cases even demand) EDI as a condition of doing business.

The Positive Effects of Adopting EDI

The industries that have adopted EDI have obtained outstanding results and proven, beyond the shadow of a doubt, that working to implement EDI is a worthy cause. The grocery industry, in particular, made EDI an instant success with the Universal Product Code (UPC).

A producer of consumer products that implemented a pilot program in 1989 found that, within six months, nearly 40 percent of its clients were communicating electronically. In response, the company was forced to take the system out of the pilot stages and put it on line immediately. In just over a year, more than 80 percent of the company's business was conducted via EDI. A universal coding scheme for all products was the key element in the program's success.

In the medical industry, EDI reduces the number of phone calls needed to adjudicate claims, replaces extensive paperwork, and reduces administrative costs. The typical flow of information without EDI might proceed something like this:

1. A patient visits the doctor.
2. The doctor fills out information about the services and procedures performed; it is transcribed into the office computer system (data entered once).
3. The doctor's office prepares "Encounter Data" for the insurance claim (possibly using a standard form).
4. The Encounter Data are faxed, mailed, or sent via courier to a Medical Management Company or Health Management Organization (HMO).
5. The claims are sorted and cataloged. The relevant claims are entered into the computer system (data entered a second time).
6. If the service is to be compensated, a check is cut.
7. The check is mailed or delivered to the doctor's office.

8. The check is deposited.

The same process, reengineered with EDI, might be:

1. A patient visits the doctor.
2. The doctor fills out a standard form.
3. Information is transcribed into a computer system where Encounter Data are extracted and sent directly to the management agent or HMO (data entered once).
4. The claims are sorted and cataloged online.
5. If the service is to be compensated, an electronic transfer automatically compensates the doctor's account.

The end result of automating the claims process, according to Joseph Brophy, is vastly improved health care services and quite literally *billions* of dollars in savings.

Common EDI Computing Platforms

EDI systems are available for almost any platform, from mainframes and minicomputers to UNIX-based workstations and PCs. PC-based EDI implementations are becoming increasingly popular, as PCs are fairly standard, are widely available, and provide a low-cost entry into EDI. In addition, the processing power of personal computers has increased tremendously in the past few years. Common PC-based EDI systems are capable of supporting 3,000 transactions per hour. High-end PC-based applications servers based on the Intel 486/50 MHz can support 22,000 transactions per hour. These are definitely not toys. Mid-range UNIX minicomputers, such as the IBM RISC System/6000 and the Hewlett-Packard HP/9000, are also very popular platforms for EDI.

The right platform, however, depends heavily on what a company has implemented already. A PC-based system that is not designed to work with a mainframe will most likely require custom programming in order to update host-based financial or inventory applications. In the instances where compatible software for the PC and the mainframe or minicomputer is not available, the price advantage of PCs is lost.

There may be tens of vendors who make specialized EDI software products for the system best suited to your application. Table 8 lists a number of vendors who produce EDI-compliant software or EDI services.

TABLE 8 Vendors of Electronic Data Interchange Products and Services

Company Name	City	State
Advanced Communications Systems	North Olmsted	OH
Advantage Systems	Waltham	MA
AIM Computer Solutions	Sterling Hgts	MI
American Business Computer	Ann Arbor	MI
Andersen Consulting	Chicago	IL
APL Group	Wilton	CT
Auto-trol Technology Corp.	Denver	CO
Automated Handling Systems	San Francisco	CA
Blue Rainbow Software International Corp.	Marietta	GA
Bonner & Moore Associates	Houston	TX
Lloyd Bush	New York	NY
Carberry Technology	Lowell	MA
Computer Associates International	Islandia	NY
Data Solutions	Wickliffe	OH
De Carlo Paternite & Associates	Independence	OH
Data 3	Santa Rosa	CA
Digit Software	Silver Spring	MD
Digital Equipment Corp. (DEC)	Maynard	MA
DNS Associates	Burlington	MA
Dunn Systems	Lincolnwood	IL
Dynamic Business Systems	Rocky River	OH
EDI	Silver Spring	MD
EDI Able	Malvern	PA
EDI Solutions	Minneapolis	MN
EDI Support	Reno	NV
EDS Canada (Electronic Data Systems)	Scarborough	ON
EMS	Manchester	NH
Engineering DataXpress	San Jose	CA
Enterprise Solutions	Westlake Village	CA

TABLE 8 Vendors of Electronic Data Interchange Products and Services (Continued)

Company Name	City	State
FORESIGHT Corp.	Dublin	OH
GE Information Services	Rockville	MD
Genzlinger Associates	Troy	MI
Grace Computer Resources	Norcross	GA
GSC Associates	Redondo Beach	CA
GTE Health Systems	Salt Lake City	UT
Harbinger*EDI Services	Atlanta	GA
Hewlett-Packard Co.	Palo Alto	CA
IBM (International Business Machines)	Armonk	NY
Imrex Computer Systems	Great Neck	NY
Information Management Consultants	McLean	VA
Integral	Walnut Creek	CA
InterCoastal Data Corp.	Carrollton	GA
Interleaf	Waltham	MA
Isocor	Los Angeles	CA
MAXXUS	San Francisco	CA
MKS	Plymouth Meeting	PA
National Systems Design Corp.	Lansing	MI
Northern Telecom	Nashville	TN
Owens Information Services	Lexington	KY
Perwill EDI	Independence	OH
Piedmont Systems	Winston-Salem	NC
Premenos Corp.	Concord	CA
Radley Corp.	Southfield	MI
RAILINC Corp.	Washington	DC
RMS (Release Management Systems)	Livonia	MI
St. Paul Software	St. Paul	MN
SDM International	Fuquay-Varina	NC
Shaffstall Corp.	Indianapolis	IN
Shared Financial Systems	Dallas	TX

TABLE 8 Vendors of Electronic Data Interchange Products and Services (Continued)

Company Name	City	State
Software Associates	Little Falls	NJ
Sterling Software	Dallas	TX
Sterling Software	Dublin	OH
Supply Tech	Ann Arbor	MI
Synergistic Systems	Neptune Beach	FL
System Software Associates	Chicago	IL
Systems Center	Reston	VA
Techview Corp.	Hoffman Estates	IL
Trinary Systems	Farmington Hills	MI
TSI International	Wilton	CT
UNISYS Corp.	Blue Bell	PA
Universal Software	Brookfield	CT
US Lynx	New York	NY
Userbase Systems	Knoxville	TN
Verimation Inc.	Rockleigh	NJ
Wellmark	Westlake Village	CA

How Do EDI-Based Systems Communicate?

Electronic Data Interchange uses some form of network to exchange data between computer systems, whether it be a telephone line, a public data network, a satellite, or a highly specialized EDI network. Smaller companies and those with low transaction volumes can use dial-up phone lines to conduct business adequately. Larger organizations may require dedicated leased lines or public data networks (typically X.25 packet-switching networks) in order to maintain the necessary level of transactions. However, organizations that are using private or public data networks must use the same software on both ends or at least the same standard for EDI. This fact makes it more difficult to share information with multiple vendors or among organizations in different industries.

Special networks called value-added networks (VANs) have been developed due to the problems with industry standards and customers' desire to use more than one vendor. VANs are third-party networks that not only take information from source to destination but also convert that information from one EDI standard to another. Most VANs use a "mailbox" to store incoming messages and forward them when the recipient connects to the network—the system is called store-and-forward. Until acceptance of national and international standards becomes prevalent, VANs may remain the logical choice for companies that do electronic business with a large number of vendors.

Current Trends Affecting Electronic Data Interchange

As with any technology-based industry, new advancements and competition often change the rules. As more and more groups begin to interconnect with EDI, various pressures from both industry and government will cause the EDI standards to change. This is one of the reasons why the use of VANs for automatically translating data between standards is attractive, even though they are more expensive than other data networks. Here are some of the issues that may affect your use of EDI in the near future:

X.12 Is Going by the Wayside

If your company is considering implementing an EDI-based system, pay close attention to the standards bodies and the particular standards for your industry. By 1997, EDI-compliant software must support the EDIFACT standard. Most United States companies are currently using X.12 or proprietary formats. Though the X.12 standard will remain, all further development of United States standards will follow the lines of EDIFACT and not X.12.

EDI Goes Global

The United Nations has gotten involved in determining worldwide standards for EDI. UNEDIFACT is a UN-approved electronic data exchange standard for administration, commerce, and trade that is meant to expedite international trade transactions. It is quickly becoming a standard format for trade transactions around the world. UNEDIFACT is being used by the United States, countries of the European Community, Japan, Indonesia, and Singapore.

UNEDIFACT has helped to standardize a list of terms and their meanings—important points in a global economy. Moreover, the EDI standard enables monetary transactions between both the exporters' and the importers' bankers. As you can imagine, this ability has greatly improved the accuracy of delivery schedules and the exchange of currency among various trading partners.

EDI and E-mail Technologies Are Coming Together

The proliferation of networks has put tremendous pressure on the computer networking industry to establish worldwide electronic mail standards. The most prevalent of these is the CCITT standard called the X.400 Message Handling System. EDI and messaging technologies are beginning to merge as the capabilities of e-mail and messaging systems improve. It makes good business sense to use the same networks for messaging of all types.

As a result, the CCITT has developed the P-EDI or X.435 standard for encapsulating EDI information into standard X.400 e-mail messages. X.435-based EDI may become more prevalent as more and more companies establish worldwide e-mail systems.

EDI Will Become the Commerce of the Future

Electronic Data Interchange is a viable technology that has been around for a long time. Although it was plagued early on by competitive pressures and lack of standards, current trends in the computing industry and the work of the standards bodies are likely to ensure that EDI will become much more cost effective and universal. As national and international standards for EDI become more pervasive and associated technologies such as e-mail and other messaging services continue to evolve, EDI is likely to become the preferred way of doing business in most industries. In the not-too-distant future, EDI will most likely become standard fare, just as credit cards, the phone system, and facsimile have become common, even ordinary, yet indispensable business tools.

Chapter 15

Mobile Computing

Mobile computing represents one of the most impactful yet sometimes least understood part of corporate computing. Many companies have neglected to build mobile computing strategies despite the fact that their work force is changing dramatically. This chapter examines the mobile computing scenario and what's available for today's dynamic organization.

In this chapter, we'll take a look at the basic types of mobile computing as they may apply to different occupations. Then, we'll look at the technologies that would be appropriate for each class of use. Clearly, mobile computing technology has progressed at light speed in the last ten years. The industry has advanced from the early luggable thirty-pound sewing machine-type computers that Compaq introduced in the early 1980s to the present pocket-sized palmtops. Today's mobile machines offer the ability to compute at any size, weight, and power. The worker can make his choice on the basis of price, capabilities, and convenience.

Defining the Mobile Computing User

The user of mobile computing technology can be the traveling professional, the salesperson, the occasional business traveler, or just about anyone who happens to go home after a day's work. Mobile computing, as we define it, could very well apply to 100 percent of the working population. That is, everyone goes home.

Generally speaking, mobile computing is reserved for the white-collar professional, most frequently the business traveler or the salesperson. Recent studies showed that well over 50 percent of mobile computer equipment was sold to traveling salespeople or traveling business professionals. What is surprising in these statistics is that the traveling business professional outranks the professional salesperson as a purchaser of mobile computing equipment. The business professional has long known that work doesn't stop just because he or she is out of the office. Now, that professional has the means to stay in touch and get the work done just about anywhere, anytime.

Technology for Today and Tomorrow

There are several different technologies included in the mobile computing market. Some are new, some have been on the market for a while, and still others are in the testing phase, destined to be the stars of tomorrow.

Portable and Laptop Computers

Portables and laptops are the original mobile machines. Compaq Computer Corporation really defined the genre more than ten years ago with the introduction of its "luggable" PC. For the first time, it was possible for a business professional to replicate his desktop computing environment while working away from the office. In the years since then, manufacturers have continued to refine the capabilities of portables while also reducing the size and price. While portable and laptop computers are still popular today, the new sales belong to the notebook market.

Notebooks and Subnotebooks

Taking the "smaller is better" trend to new heights, manufacturers introduced new mobile computers that are small enough to fit inside a briefcase. This range of portable computers is popularly known as "notebooks." These machines typically are about the width and height of a sheet of paper, with a thickness of about two inches (or less) and an overall weight of seven pounds or less. There is another class of even smaller mobile computers hitting the market known as "subnotebooks." These machines prove that good things come in small packages. The notebooks and subnotebooks hitting the market today are built around powerful CPUs (generally the Intel 80386 or 80486), and they can run complex applications such as Microsoft Windows and other productivity software.

The key benefit of notebook and subnotebook computers is the ease of taking them on the go. These devices are small and light, and yet they don't sacrifice much in the way of features and functionality; they seem to have it all.

Palmtops

Palmtop computers represent the smallest of the small. They literally rest in the palm of your hand, weighing only about two pounds. Most palmtop computers have a limited range of use, given the small size of the keyboard and monitor. For instance, you wouldn't want to construct or give a graphics presentation on a palmtop computer. Team it up with a wireless modem, however, and you have an excellent machine for electronic mail and simple file transfer.

PCMCIA Technology for Peripherals

A relatively new consortium called the Personal Computer Memory Card International Association (PCMCIA) is redefining the way we think of mobile computing. Products built on the PCMCIA standards are miniaturized versions of the peripherals that we consider essential: modems, hard disks, memory cards, LAN adapters, and more. They are being incorporated into the computing platforms we've just mentioned—portables, laptops, notebooks, subnotebooks, and even palmtops.

PCMCIA cards are about the size of a credit card but slightly thicker. Don't let the size fool you—these cards pack a wallop. The incredibly small size of these cards makes mobile computing more convenient than ever. The earliest cards typically provided additional memory or modem communication. Looming on the horizon are a host of peripherals, such as sound cards, CD-ROM controllers, wireless LAN adapters, and tape backup drives. What's more, with new notebook computers having multiple PCMCIA slots, it's possible to reconfigure your computer virtually on the fly. PCMCIA addicts have taken to carrying peripheral cards as they would credit cards—in a leather case resembling a wallet.

The prices for PCMCIA products are relatively high today, but that situation won't last long as more products come to market and competition drives prices down. Reliability and compatibility are steadily improving as well.

Pen-Based Computing

Pen-based computing is not new; it has been around for at least five years. It hasn't yet reached critical mass, however, largely because the technology is still evolving. Operating system standards are yet to be determined, with the front running systems coming from Microsoft Corporation (Windows for Pen), Go Corporation (PenPoint), and Communication Intelligence Corporation (PenDOS). Reliable and consistent handwriting recognition continues to be a stumbling block. Moreover, there is a dearth of commercial applications at this point. Most of the available applications are for vertical markets.

Despite the obstacles, we hold out hope for pen-based computing, particularly for keyboard-phobic people. Think about it. This is a technology that allows people to interact with a complex computer in a very simple and comfortable way: with the written word. Pen-based computing can open the door to computing for people who can't tolerate the inconvenience of a keyboard. As an example, look what pen computers have done for the package delivery industry. UPS drivers now get recipients to sign for packages electronically, eliminating the need for cumbersome paper forms and redundant data entry.

Personal Computing Redefined

Another new technology that is just on the horizon is what has come to be known as personal digital assistants (PDAs), personal intelligent communicators (PICs), or mobile companions. These handy all-in-one devices promise to combine several voice and data functions in compact, portable packages. Eventually, they should replace your subnotebook computer, your cellular phone, and your wireless modem. Applications will run the gamut from an electronic Day Timer and personal scheduler to a communications center.

Wireless Computing

One of the newer technologies to further the cause of mobile computing is wireless communications. With wireless, messages and data files are transmitted via radio signals. Wireless communication can be used for simple applications, such as electronic mail, or for remote access to a homebound LAN.

Staying in Touch: Remote Access

It's nice to take your computing environment with you, but what happens when the data or applications that you need are located on your desktop PC or on the departmental LAN back in the office? Happily, your regular files and applications on your office computers are just a phone call away.

There are several options for remote file retrieval or application use. One access method allows you to set up your desktop PC in a host mode, which permits you to send and receive files and leave messages, just as you would with a bulletin board system. Another method is essentially a remote control program. When you dial in to your deskbound PC, you take complete control over it, just as if you were sitting right there at the keyboard. This solution permits you to access entire applications, not just files and messages. It vastly increases the capabilities that your mobile computer offers.

There are also products that provide you with remote access to the network. You can dial in to your host desktop PC and log into the LAN from wherever you are. Better yet, you can dial into the LAN directly via a network modem. Each option has its conveniences and drawbacks, but the main benefit is that you can have easy access to your office computers without physically being there.

Something's in the Air

One of our favorite uses of wireless technology is the transmission of electronic mail messages. RAM Mobile Data and Ericsson/GE have collaborated to bring to market a complete wireless e-mail solution called RadioMail. The package, called Viking Express, includes a Hewlett-Packard palmtop computer and access to the airwaves.

Since it is e-mail, the system offers all the advantages of wire-based electronic mail, without the hassles of being deskbound. Mail can be sent to multiple people simultaneously; there is a permanent record of the message that is transmitted; mail can be sent or read whenever it is convenient; messages can be forwarded to other people. Compare this to cellular phone technology, where inbound calls can't be controlled and calls can't be stored or forwarded.

The Currid & Company team has become absolutely addicted to this fast, efficient, and inexpensive method for sending and receiving messages virtually anywhere and anytime. Because it offers the benefit of near-real-time communication, we can stay in touch without being tethered to an RJ-11 jack or even to a telephone.

The Outlook on Mobile Computing

A recent Datamation/Cowen & Company survey indicates a sharp rise in purchases of notebook computers. Moreover, the survey respondents expressed more interest in PC/notebook fax software than in workgroup computing, multimedia, or pen-based computing. Still, the interest in mobile computing has yet to reach its peak.

Mark Eppley, CEO of portable communications vendor Traveling Software Inc., attributes the slow rise to the peak to the general immaturity of the mobile computing market. Says Eppley, "We're at about the equivalent of the 1915 auto market" in terms of mobile computing's refinement and ease of use. The two major limiting factors are noise in cellular technology (limiting data transfer rates) and short battery life (limiting the amount of work that can be done). Eppley predicts that "electronic filling stations" will emerge to alleviate these problems. The filling stations could be located in such places as airports and convention centers. Users could "plug in and fill up" their devices with data from their own enterprises via public data access networks. They could also download relevant information from commercial data providers.

Other proponents of mobile computing argue that our entire society will be transformed by the marriage of wireless communications and mobile computing. Motorola Inc. executive vice president and general manager Robert Growney says, "There is an enormous appetite among voice and data users to become untethered." Industry giant IBM is throwing its support behind the push for mobile computing. In fact, the joke at IBM is that the company acronym stands for "I Believe in Mobile." Apple Computer, Inc., is also jumping on the bandwagon with its as-yet-unreleased products for the personal communications services (PCS) market. Apple chairman John Sculley predicts that PCS-related products and services could create a $200 billion marketplace in the next decade.

What Does It Mean to You?

Are you still clinging to a shred of doubt that mobile computing has real benefits for the corporate world? We'll give you two actual examples of how mobile computing proved to be a real boon to Currid & Company's daily business.

The author, Cheryl Currid, company founder and president, is a frequent business traveler. In some months, I can spend more time in the air and on the road than I spend at home. I am mindful of how much time and money are wasted on "dead space"—time that is spent in travel that normally couldn't be used to transact business. For example, the cab ride from the airport to the hotel is usually dead space. The time spent in line waiting for a rental car is dead space.

During one of my recent trips, I used my time on the plane to type out a bunch of messages on my wireless e-mail system. Since wireless computers can't be used on planes (due to interference concerns), the messages were stored until the plane landed. Once on the ground, I released 14 new messages into the air. By the time I picked up my luggage, I had received 8 messages back. While standing in line at the rental car counter, I read and responded to my new messages. I was able to conduct business as if I were sitting at my office PC, making effective use of what used to be dead space.

On another of my road trips, I managed to update a magazine column at the editor's request and have it turned in to meet a deadline. I was in New York, and the editor was in San Francisco, waiting to board a flight for New York. He sent me e-mail asking for slight changes to my column. He attached an electronic copy of my column to the message. I received the message with the file over my wireless e-mail network. I saved the file to a PCMCIA hard-drive card in order to transfer the file from my palmtop to my larger notebook computer. I edited the file, saved it out to the PCMCIA card, transferred it back to the palmtop PC, and included the updated file in a new e-mail message. By the time the editor stepped off the plane in New York, he had received my column updates. He, in turn, forwarded the file to the magazine, which received the column in time for the deadline.

If you aren't impressed by these stories, stop and think what the scenarios would have been without mobile computing technology. In the first case, my time on the plane and in the car line would have been wasted. Furthermore, I would have been out of touch with my e-mail correspondents until my return to the office days later. That could have meant lost business opportunities. In the second case, I would have had to dictate my column changes over the phone—that is, assuming my editor could even reach me. Remember, he found me through wireless e-mail in the first place. Now you can begin to see the impact that mobile computing can have on a business. The traveling professional can cheat the clock and recoup what once was lost time.

The State-of-the-Art Machine for the Road Warrior

Businesses often achieve their competitive advantages by using key resources to the fullest. Many times, those key resources are the employees themselves. That's why prudent companies see the benefit in equipping their frequent travelers with the best possible machine for mobile computing. With today's technology, it's possible to tote a computer with all of the power and features of your desktop PC, yet small enough and light enough to fit in your briefcase or on an airplane tray table. What's more, you can have it all typically for less than $5,000.

This section describes what we feel is the ultimate machine for business travelers. We limit our scope to notebook and subnotebook computers, since size is an important factor when one is traveling a lot.

When shopping for the road warrior's winner, be sure to select a machine that's long on CPU power but mindful of energy needs. In our opinion, it's barely worth looking at anything less than a 486-based PC, although older-model 386-based notebooks can meet many needs at a low price. If you plan on working with graphics at all, go for the notebook with the higher-powered CPU.

Energy consumption is not generally a major concern, although notebooks can vary widely in their energy features. At a minimum, you want a machine that will give you several hours of reliable use on a rechargeable battery. Most notebooks also have a separate AC adapter for those times when an electrical outlet is available. Manufacturers of portable computers tend to equip their machines with all sorts of energy-saving features, such as "sleep" modes. Such features will become more popular (and even mandatory) as new government guidelines for energy consumption go into effect under the Energy Star program.

Hand in hand with a powerful CPU is an adequate amount of random access memory (RAM). We recommend a minimum of 4 megabytes but prefer more if possible. Applications like Microsoft Windows are memory intensive, and no one wants to wait ten minutes just to access an application.

After speed and power, perhaps the next most important feature of a notebook computer is a comfortable keyboard and pointing device. What is comfortable? Well, that's a matter of taste. Most people, however, want a keyboard that closely matches the one on their desktop PC. No one wants to search for keys or press weird key combinations like SHIFT+CTRL+F1 just to move the cursor. Moreover, the keys should be large enough and spaced well enough to accommodate the typical adult-sized hand comfortably.

Pointing devices are another important feature, especially for GUI users. Most notebooks today have built-in devices, with a port to accommodate a separate mouse if desired. Some manufacturers have come up with truly innovative ways to incorporate a pointing device. IBM, for example, integrates a pointing controller called the Trackpoint II into its ThinkPad notebook. The Trackpoint looks like a pencil eraser stuck in the middle of the keyboard, and it can be accessed easily without removing your fingers from the home row of keys. Compaq has implemented a thumb-driven trackball on the display screen of its notebooks. The Hewlett-Packard Omnibook 300 has a slick little slide-out mouse to the right of the keyboard. Right-handed people should find it easy to use. Sadly, though, lefties are left out by this device.

As for our display, nothing less than color will do. Color adds a significant dimension to the computer, making it possible to distinguish more information and icons. The best of today's notebook technology is the active matrix display, and the larger the screen, the better. Also, make sure the machine has a port to accommodate an external monitor.

Next look for a big hard drive—at least 120 megabytes. Today's software applications are using more and more disk space, sometimes as much as ten megabytes for just one application. Allow yourself plenty of disk space for applications and data.

Communications is the key application that most business travelers need. The research group Market Intelligence predicts that sales of portable high-speed modems will quadruple by 1998. So, your ultimate notebook must have a built-in (or convenient plug-in) high-speed modem; 14,400 bits per second is the current preferred speed. Fax modems are gaining in popularity, too.

And now for the extras that make this machine "the ultimate." Select a machine with PCMCIA slots. Even though there aren't many PCMCIA peripheral products available today, this is the wave of the future. Make sure your machine can accommodate the cards as they do become available. Other niceties include battery gauges and removable, upgradable hard drives. Some machines, such as the Apple Powerbook, fit into a desktop expansion base to turn your portable notebook computer into a desktop dynamo. You essentially get two machines for a little more than the regular price of just a desktop system.

If printing is important to you, there are plenty of "notebook"-sized printers available. Older portable printers left a lot to be desired in terms of output quality. Today, however, you can get PostScript documents on a 300 dots-per-inch printer that's small enough to fit into your glove compartment.

With a notebook computer equipped as we have specified, the traveling professional can meet just about every computing need while he's on the road.

Back to the Future?

We firmly believe that mobile computing technology will redefine the work environment. More importantly, it opens up options for alternate work styles and the ability to work whenever and wherever people choose. The technology we discussed in this chapter and the uses we see today constitute just the tip of the iceberg.

Mobile computing technology—taking the form of small computing and communicating units—represents both sociological and technological breakthroughs. For the most part, the technology is inexpensive to buy and can be brought to organizations without major upheaval. Smart organizations will investigate and make good use of this technology.

Chapter 16

Imaging, Scanning, and Multimedia

Imagine sitting in your living room watching the Sunday Night Mystery and helping the detective discover who killed Mrs. Jones. Then, before you go to bed, you watch a customized news program with items of personal interest to you preselected. Upstairs, your teenager is accessing the Library of Congress, conducting research for a school report. The next day, you convene a team of experts from around the world to put together a proposal for a key client, without anyone having to leave his office. On your way home, you stop by the local music store and download a custom CD of your favorite songs. Sound far fetched? Well, maybe—there are still several technical hurdles to clear in order to provide these kinds of services. However, the future is not that far away.

The entertainment industry has long used advanced technologies to produce the awe-inspiring special effects you see in box-office hits like "Terminator 2" and "Jurassic Park." Their use of inexpensive computers to produce dazzling special effects was just the beginning of a new wave of applications for technology. The entertainment industry has led the way to the full-featured interactive media systems now under development. At stake is the heart and mind of the consumer and the 21st-century knowledge worker. Fundamentally, interactive media have the capability to fundamentally change the way we work, play, and learn. In addition, potential revenues are well into the multi-billions. This is the primary reason so many big players in the entertainment and high technology industries are so interested in this technology.

If you can relate to any of the scenarios that opened this chapter, take heart, there is a whole new wave of technology about to invade every aspect of your life, from your leisure activities to your work environment. In this chapter, we'll discuss the events and catalyst technologies that will make for this revolutionary change in our personal and business lives. We will also discuss how you can take advantage of them to provide a competitive advantage for your organization.

Multimedia

The term multimedia has been used to loosely define a variety of applications. It encompasses a broad range of dissimilar applications of media integration. It includes voice or other sound, video, graphics, and text. Media integration refers to use of various forms of information to improve the effectiveness of the message. Figure 8 defines the types of information and their states.

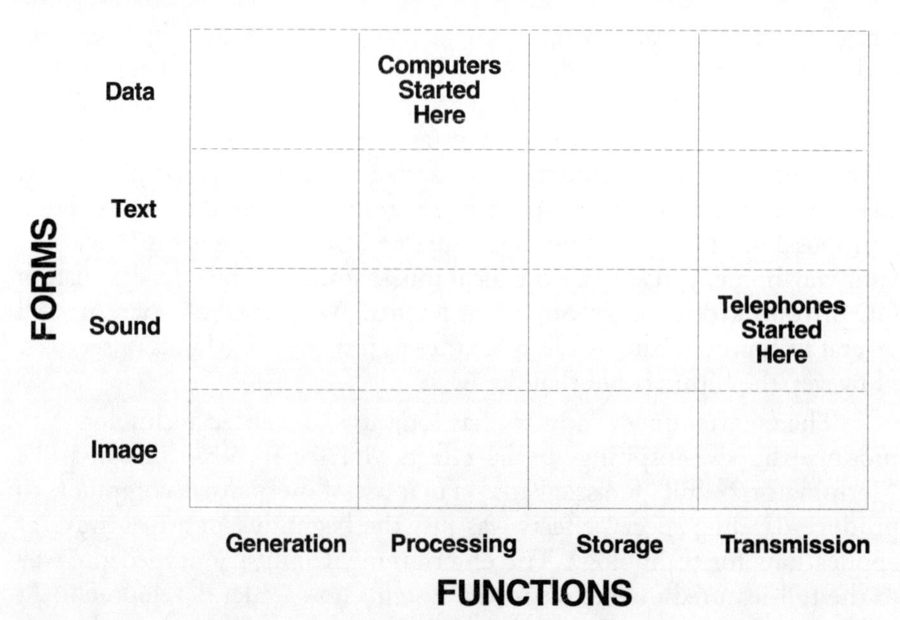

FIGURE 8 Architecture of information

Stan Davis and Bill Davidson, in their book *20-20 Vision,* describe the meaning of each region of the grid in the figure as follows:

Forms

The rows in the figure represent the forms of the original data, corresponding to the media that they come from:

- Data: Facts, numbers, letters, symbols, and the like that can be processed or produced by a computer. Obviously, data existed before the invention of the computer, but it is the computer's ability to handle large quantities and volumes of data that is significant.
- Text: Written language, either in printed or handwritten form. What is important here is that, as technology progresses through the century, it will begin to handle and interpret handwriting. This ability has been commonly referred to as electronic ink. Its lack is why computers have never really replaced the fax machine. In fact, most companies have significantly increased the use of faxes over the last year.
- Sound: What we hear, with an emphasis on music and the spoken word.
- Images: Visual forms of information, such as drawings, illustrations, and photographs.

Functions

The columns in the figure represent the functions applied to each form—the successive transformations that data from each original medium undergo:

- Generation: The first of four functions by which information in one of the four forms described above is processed. Generation refers to initial capture of the information in digital form.
- Processing: Computes, converts, edits, analyzes, and synthesizes information.
- Storage: Stores information for later use.
- Transmission: The final function, sending and receiving of all forms of information.

The Prime Drivers

Multimedia technologies are being driven by the merger of four key industries: 1) entertainment; 2) the computer industry; 3) communications; and 4) publishing, as shown in Figure 9.

All these industries are necessary in order to realize the true potential of multimedia. The application possibilities produced by this collaboration are truly limitless. Industry efforts are basically focused on two primary areas: consumer services and business applications. Each area will use the same underlying technology, but each has a different objective. In the first case, it is to usher in a new age of personal entertainment, and, in the second, it is to help shape and mold the virtual corporation.

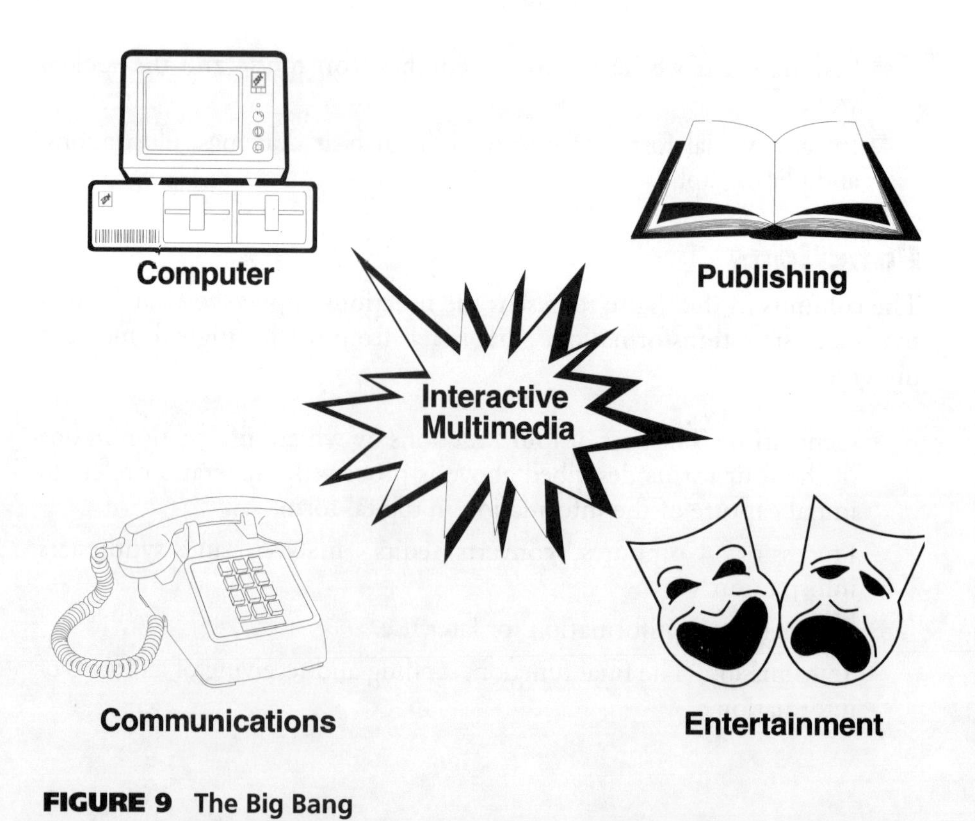

FIGURE 9 The Big Bang

Interactive Media

The current environment recalls sharks in a feeding frenzy. If you lower a truckload of raw meat into shark-infested waters, the sharks will go into a feeding frenzy, eating everything in sight, including each other. When it is all over, only a few survive. We are currently in the middle of the interactive media feeding frenzy. Table 9 is a brief chronology of relevant events over the last few months.

TABLE 9 Chronology of Interactive Media

December 1992	Tele-communication, Inc. (TCI)	Announces that it will use digital technology developed by General Instrument to provide 500 channels to its cable subscribers.
January 1993	AT&T, MCA, and Time Warner	Invest 15 million in 3DO, a start-up company that is building an interactive disc player scheduled to be released by Christmas 1993.
February 1993	Southwestern Bell	Buys two cable systems in Washington, D.C. from Hauser Communications.
April 1993	Sega, Time Warner, and TCI	Announce that the Sega Channel will be available by 1994.
April 1993	Microsoft, Intel, and General Instrument	Reach an agreement to incorporate personal computing power into cable TV converters.
May 1993	U.S. West	Buys a quarter of Time Warner.
May 1993	AT&T	Announces a video server that will allow consumers to select movies on demand from the comfort of their own home.
June 1993	Macy's	Announces that it will enter the home shopping market.

At this point, you are probably wondering what interactive media have to do with the business environment. At first glance, it may not seem very obvious. The events shaping the entertainment landscape, however, will have a dramatic effect on your business for two reasons. First, technology initially developed for entertainment use will be adapted to business applications, and technology initially developed for the business environment will be adapted to entertainment applications. A good illustration of this point is the cross utilization of interactive media technologies. For example, Japanese businessmen are able to interact with the stock market from home using a Nintendo Entertainment System. Video technologies for computer monitors have become the basis for digital High-Definition TV. Second, giants in their respective industries (e.g., Nintendo, AT&T, IBM, Apple) all have their eyes on expanding beyond their current market share. Entertainment companies are looking to provide business services, and traditionally business-focused companies are positioning themselves to enter the entertainment market. The fact of the matter is that the lines between the two are beginning to blur, and new words like "infotainment" or "edutainment" are beginning to emerge. These describe the combining of information, education, and entertainment.

Applications

The idea that information can take many forms and exist in many states isn't new. What is important, however, is that technology has finally reached a point that systems are able to handle the complete grid (as shown earlier in Figure 8) at an economical price-performance ratio. It is this ability that opens the way to an explosion in the development of new and innovative applications designed to accomplish one task—to turn information into knowledge so that an individual can act upon that knowledge. The result is that a whole new wealth of information will be available to the individual. For the most part, many high-end desktop computer systems sold today have the capability to support the full range of multimedia. Several years ago, this would not have been the case. A multimedia-capable system would have been priced well beyond the purchasing power of the average user.

The following is the set of key technologies and applications based upon leveraging the full range of information forms and functions.

Teleconferencing

A teleconference may take several basic forms. The first is the video teleconference, today usually conducted in a room especially configured for the purpose. Over the last few years, we have seen the prices of these systems drop dramatically, from around $100,000 to around $20,000. In addition, the amount of bandwidth required to support a video teleconference has decreased significantly. Eventually, video teleconferencing will make the transition from the conference room to the desktop. The second form of teleconferencing is voice teleconferencing, which is commonplace and available through most business phone systems. In addition, it is even possible for a home telephone subscriber to purchase a conference calling option that allows her to conduct a three-way voice teleconference.

The final form of teleconferencing is computer teleconferencing. We will discuss this form in more detail in Chapter 17, which deals with groupware.

Electronic Publishing

With the advent of the compact disk, electronic books have become a very common commodity. Although the initial attempts are very rudimentary, they do display the promise of the technology. A typical CD-ROM (compact disc, read-only memory) can store as much as 660 megabytes, with the systems' capacity increasing monthly. The large storage capacities allow publishers to combine the complete range of voice, data, video, and imaging, providing a very productive learning environment. CD-ROM storage is becoming very popular in large corporations that need to archive financial statements, since CD-ROM disks, being proof against modification, offer greater security. The compact disk will also become invaluable to research departments because of their large storage capacity.

Multimedia Electronic Mail

Multimedia e-mail is supercharged mail, which will allow people to incorporate voice, data, video, and images into an electronic message. The real advantage to this form of communication is that, in allowing incorporation of various media into messages, it significantly improves the quality of information exchanged by individuals.

Imaging

Imaging refers to the process of scanning paper documents into electronic pictures for online retrieval and processing. If your organization seems to be drowning in paper, imaging may help. Several innovative companies have used imaging technologies to increase individual productivity dramatically and to improve customer satisfaction. Imaging, when properly deployed, can cut overall costs and improve overall quality. For example, many companies have found that imaging helps them keep track of work in progress more efficiently. Workers don't have to thumb through file cabinets to find a client's records. Rather, information is instantly available to anyone who needs it at the touch of a few key strokes.

One insurance and financial company that has been a pioneer in imaging technology is USAA of San Antonio, Texas. Upon his arrival in 1968, Robert F. McDermott, chairman and CEO of USAA, recognized that the company had a problem dealing with customers, and that the problem arose from the company's inability to handle the paper log effectively. Paper was everywhere, and misfiled or missing documents were commonplace. At the time, USAA had only about 650,000 members. McDermott realized that, if the company was to handle the increasing demand for its services, something would have to change. As a result, the company initiated a $100 million systems strategy to improve overall responsiveness and customer satisfaction. Imaging turned out to be a key component of that strategy. Today, USAA has grown to over 2 million members and has established itself as a leader in the insurance industry because of its outstanding customer service. The key to the organization's success was complete reengineering of all internal processes along with implementation of strategic technologies such as imaging.

Scanning

Scanning refers to the process of reading text, images, and bar codes and then converting them into an electronic digital image or code. Graphical scanners are a key component to a document imaging system. It is important to note that graphical scanners do not recognize any of the actual content of the text or image. They just convert paper to a digital image. On the other hand, bar code scanners recognize printed fonts or symbols and convert them into machine-readable code. If you've stopped by a grocery store lately, you have probably seen a bar code reader at the checkout line. Overall, bar code readers have significantly improved productivity of individual grocery store clerks. More importantly, bar code readers allow a store to collect consumer information at the point of sale. This information allows individual stores to determine the best product mix as well as store layout based on actual consumer preference. National suppliers are able to utilize this information to take into account regional preferences. This process is a perfect example of mass production customizing.

Fiber Optics

The term *fiber optics* refers to a technology in which light is transmitted through a fiber cable to transport information from one point to another. Fiber optics can transport data much faster than conventional copper cable and is currently the transmission medium of choice to build the national information superhighway. Fiber optics is also a very strategic technology, in that it can provide the transmission capacity to carry the complete information grid.

Compression

Compression techniques reduce the representation of the information in its electronic form. There are two basic types of compression techniques. The first type is non-destructive compression, which does not change the information content. This technique is primarily used for data compression applications. The second type is destructive compression, which actually degrades the quality of the original signal or information content, but which allows greater reduction of the original. This technique is primarily used for applications that can tolerate bit loss such as video teleconferencing. The key advantage of compression as such is that data will take less time to transmit and less storage capacity.

Cyberspace

Cyberspace is a term coined by William Gibson in his novel *Neuromancer.* It refers to a futuristic network that people use by plugging their brains into it. I don't know if we will ever get to a point where we will literally plug our brain into a network. However, the real-world cyberspace is the electronic realm where people do plug in their intellects.

HDTV

HDTV stands for High-Definition Television. Today's typical TV set in North America contains 336,000 pixels. An HDTV set has the same resolution as a motion picture or 35mm slide, providing at least two million pixels of resolution. In addition, HDTV as deployed in North America will be digital, unlike today's analog system.

Virtual Reality

Virtual reality is a synthesis of computer-generated stimuli that projects the user into a 3-D space. It involves the user's immersion in and interaction with another world, one that simulates real-world situations and environments.

Visualization

Visualization systems combine computerized graphics and imaging technology and present the result in a visual form on a workstation or personal computer.

"Informationalized" Commodities

As processing power continues to drop, the most significant result of the new technologies is that everyday commodities will be transformed through the creative use of multimedia. In other words, they will become "informationalized," allowing suppliers to customize products and services to the individual consumer. Some good illustrations in our everyday life include automatic teller machines, information kiosks (such as the map and direction system at a Hertz Rental Car counter), and restroom fixtures with infrared sensors that turn faucets on and off automatically. The question you must answer is: How can you "informationalize" your current product line?

Will Multimedia Make It and Will We See It in Our Lifetime?

The real question is not whether the technology will make it but when. Sooner or later, unquestionably, multimedia systems and applications will become as common as the telephone. The question remains, when will consumer demand cause this to actually happen.

In her book *The Popcorn Report,* Faith Popcorn identifies 10 trends that will be prevalent throughout the decade:

1. Cocooning
2. Fantasy Adventure
3. Small Indulgences
4. Egonomics
5. Cashing Out
6. Down-Aging
7. Staying Alive
8. The Vigilante Consumer
9. 99 Lives
10. S.O.S. Save Our Society

According to her, for a product or service to be successful, it has to meet at least three of these trends. Well then, hold on to your hat; the interactive media concept is a lot more than hype, because, in our estimation, it meets at least seven of these trends:

1. Cocooning: Basically, cocooning is the impulse to go inside when it gets just too tough outside. Multimedia allows us to do this in a variety of ways. Technology has allowed the cocoon to move from being a place to a state of mind.
2. Fantasy Adventure: A fantasy adventure is a vicarious escape through consumerism, catharsis through consumption. Send us to a far-away place to fulfill our heart's desires without any risk, but we want to be back in our own bed by 10:00 P.M. Although we have along way to go to get to the holodeck aboard the Starship Enterprise, interactive media are a definite step in that direction.
3. Egonomics: Egonomics basically caters to a consumer's need for personalization. Technology will allow goods and service providers to tailor their products to the smallest market niche possible—the individual consumer. You can already go to some hair stylists and sample a new hair style without touching your real hair. In Japan, cars are customized for individual consumers. There really isn't such a thing as a car showroom. Be ready to be subjected to a barrage of these types of services direct from the comfort of your living room. If you are in the business of selling a product or service (which most of us are), you should be mapping a strategy to modify your marketing approach to take advantage of this trend.
4. Cashing Out: Over the last ten years, we have witnessed a total collapse of the corporate myth. We no longer believe in the parent corporation, simply because corporations haven't really been very good parents. In today's environment, loyalty doesn't produce job security. As a result, more and more people are cashing out and heading for the hills. Interactive multimedia technology and rich access to information have provided the tools that allow people to live and work where they choose. Has the time come for your company to revisit telecommuting?
5. Down-Aging: Nobody wants to get old or to feel old. We will embrace anything that helps us feel young. Techo-toys will do this. How many parents do you think play Nintendo as much as their kids?—Probably many more than will admit it.
6. The Vigilante Consumer: The new wave of multimedia and interactive media will produce a very knowledgeable and activist consumer who is willing to fight the corporate power mystique, to right what he

feels to be a wrong. If you underestimate the importance of instant, accurate information to the consumer, just think back to the November 1992 presidential election. The landscape of politics was changed forever. More voters tuned in and turned out than at any other time in our history. However, the process didn't stop there. In fact, John Q. Public has made a conscientious effort to stay informed and stay involved. Every move President Clinton makes, he makes under John's watchful eye. Little slips are no longer tolerated, and the public (the consumers) want to get what they voted for without the usual pile of propaganda.

To sum up: Deliver or die. Customers will be increasingly demanding. Multimedia technologies can help you track your customer preferences, goals, or demographics in order to improve overall quality and responsiveness.

7. 99 lives: Simply stated, we want our time back. We are being inundated with information from a variety of sources, most of it useless to the receiver. The new wave of technology will help filter information, allowing us to pick and choose the information we need automatically rather than manually. In other words, as an employer, you will be able to increase workforce productivity, while still giving employees time to enjoy life.

In our opinion, multimedia technologies don't really meet the remaining three trends, labeled "small indulgences," "staying alive," and "save our society." "Small indulgences" refer to the little things in life we do to reward ourselves without worrying about the consequences. For instance, you may decide to treat yourself to a steak and lobster dinner because you finally finished an important project. "Staying alive" refers to our obsessive quest for a healthier and better life. Although you may stretch the definition to include high-tech exercise equipment, on the whole, multimedia technology really doesn't meet this trend. Finally, the "save our society" trend refers to an effort to make the '90s a socially responsible decade focused on the three critical E's: Environment, Education, and Ethics. Although multimedia technologies may be used as a tool to aid organizations in taking advantage of this trend, it is important to keep in mind that it will play only a secondary role, if any.

Key Players

It's hard to identify any one player that will have the primary role in the ultimate success of these technologies. We will see a rash of mergers and strategic partnerships develop. No one company can do it alone. Alliances between computer industry companies will drive the market, even if most of the innovation will spring from small companies. The companies identified in Table 10, however, will be the key players in the evolution of this technology and the development of a new industry.

TABLE 10 Key Players in Multimedia

Computer Industry	Communications Industry	Entertainment Industry
IBM	RBOCs	Nintendo
Microsoft	AT&T	Sega
Apple	Cable TV Industry	Sony

Where Multimedia Fits

Multimedia, scanning, and imaging will not solve your information nightmare, which will probably get worse before it gets any better. These technologies are, however, a step in the right direction. In the long run, interactive media technology will do for information assimilation what the printing press did for publication. The question is, Will society be ready for it? Is your company ready for it?

Chapter 17

Groupware

Today's organization has trimmed many management layers and flattened the overall corporate hierarchical structure. This move cuts through organizational boundaries to empower individual employees as well as to promote an environment where ad hoc workgroups can be formed to accomplish a specific task. The issue most organizations have to overcome is that individual workgroup members don't necessarily work in the same locations. In fact, as organizations continue to spread across the globe, it is very likely that a workgroup could span several continents. Enter the new wave of office productivity software—groupware.

What exactly is groupware? Well, if you ask five different people, you will probably get five different answers, because the term "groupware" has been applied to a variety of different products and concepts. Briefly put, groupware consists of applications designed to improve and facilitate workgroup interaction and collaboration. This chapter will define the various components of what many people have termed groupware as well as give you an understanding of how groupware may be applied within your organization to boost workgroup productivity.

Groupware Defined

Today's LAN environment was developed for one purpose—sharing: sharing printers, sharing disk storage, and sharing communications gateways.

Groupware adopts this fundamental principle of LANs, builds upon it, and then takes it to a whole new level: A level where users are free to forget about the underlying technical architecture. A level where users demand more out of their systems than individual productivity tools can provide. A level where users want to share ideas and information. Groupware is about human-to-human communications.

Groupware has three basic components: information sharing, messaging, and collaboration.

Information Sharing

Information sharing is an evolutionary step from today's shared database environment. The underlying principles are basically the same, but the how, what, and why has changed. The fundamental principle of information sharing in a workgroup environment is to improve and build upon team memory. Broad access to information is critical to the success of any highly efficient workgroup. The more the entire group knows, the more likely that individual efforts will remain synchronized.

The information-sharing component can be very critical to an organization's work-flow design effort. I'm not talking about just automating a specific task. What I am talking about is using free access to information to change the underlying process. Some tools that can be used to support an information-sharing environment are as follows:

Work-flow Design and Management Systems

Work-flow design and management systems are designed to mirror and model your organization's processes and procedures. You can almost classify them as "thing" trackers. A robust work-flow design and management system will be flexible enough to manage your help desk, track defects, and then order lunch for the office.

Scheduling

If you have ever tried to coordinate a meeting among five very busy people, you can relate to the need for a group scheduling system. Group scheduling systems allow you to check everyone's schedule, resolve conflicts, find a conference room, and recommend the best time and place for the meeting to occur. Sounds simple, but the underlying technology to support a group scheduling algorithm can be very complicated.

Document Management

We briefly touched upon document management in the course of the discussion of imaging in Chapter 16. Document management systems allow workgroups to place a library of information on line so that anyone can access the information at any time.

Shared Databases

The underlying information engine required for a robust information-sharing environment goes well beyond traditional database architectures. The information engine must support and manipulate more than alpha-numeric data. We have progressed beyond this point to where we want the database to hold images, graphics, voice, video, and about anything else we can think of that has an information content.

In addition, the system must support multiple versions of "work in progress," that is, copies of a document opened by more than one individual at a time, while at the same time maintaining the integrity of the overall document. The trick is consolidating the various pieces back into one logical piece of information. Representative applications include project management systems and contact management systems.

Messaging

In today's office environment, messaging occurs in a variety of media: voice mail, Post-it™ notes, fax, and electronic mail. However, it is the last item, electronic mail, that has become the foundation for groupware applications designed for exchange of information among workgroup members.

E-mail is becoming the cornerstone for groupware applications because it provides the infrastructure that allows not only individuals but applications to communicate. Once upon a time, e-mail was used only by a handful of technicians and scientists. Things have changed. E-mail is now just as much a necessity as the network operating system. In fact, it is growing very difficult to distinguish between e-mail as an application and e-mail as an operating system service. The boundary is really turning gray. For instance, a whole new wave of applications has been released that are "mail enabled." In short, these applications use e-mail systems to transport, store, and forward messages as well as to provide users with access to electronic mail from within an application.

One key point to keep in mind about e-mail as your basic message transport system is the fact that the standards have yet to reconcile various vendor implementations, such that any one system can interpret a message from any other perfectly. One way to get around this problem is to implement a messaging gateway built upon the X.400 standard. These gateways will take a proprietary formatted message, convert it to a standard X.400 message, transport it over the backbone network, and then translate it into the recipient's proprietary format.

Collaboration

The last component of a groupware system is collaboration. This is probably the least understood of the three, simply because it is the hardest to define and the newest in terms of application support. It is, nonetheless, the most important component of the overall groupware architecture.

Collaboration tools help groups think and work together whether they are in the same room or several thousand miles apart. Collaboration tools come in a variety of shapes and sizes, all designed to promote various forms of human interaction. There is really no one right way to build and use a collaboration system. The fact of the matter is that different workgroups will work differently, and they must have the flexibility to use tools on an ad hoc basis in order to achieve maximum effectiveness. A few examples of collaboration tools are as follows:

- Brainstorming: Brainstorming software allows participants to brainstorm various aspects of a given project or idea collectively.
- Idea Organization: Idea organization software condenses the ideas that result from a brainstorming session into a set of relevant topics.
- Group Editor: Group editors allow individuals to contribute sections to a working document by commenting on other people's work without altering it.
- Video Teleconferencing: Video teleconferencing systems support real-time transmission of voice and video signals. Today's systems are primarily conference-room oriented. Although these types of system will remain in place, desktop video teleconferencing systems will begin to fill a void in personal communications on an ad hoc basis. They will also support non-real time communications in the form of electronic video notes and files.

Groupware Objectives

The primary objective of a groupware application should be to aid workgroups in thinking, communicating, and working together regardless of location. To do this, the application has to support various aspects of formal and informal communications that include:

- Same Time Same Place: This can be considered the normal office environment. John, Becky, and Joe all share the same office and work space. Chance meetings in the hall are commonplace, ad hoc meetings are a normal occurrence, and group interaction is very dynamic.
- Same Time Different Place: This situation arises when two groups or individuals need to communicate instantaneously but they are physically located in different places.
- Different Time Same Place: We have all needed to communicate with a coworker, but found that coworker temporarily out of the office or in the middle of another task.
- Different Time Different Place: In our global economy, this is becoming the norm rather than the exception.

Market Leaders

Most of the major operating system vendors and software development companies recognize that groupware applications represent the next major productivity gains in personal computing. Consequently, each has taken a different strategy in providing solutions to meet the changing demands of this emerging market. This section will discuss the key initiatives for each of the major suppliers in this market.

Apple Computer

Apple touts its System 7 operating system as the basis for its groupware offering. The primary selling point is System 7's ability to share files, folders, and information formats without a dedicated server.

Banyan

Banyan's initial offering focuses on providing an intelligent message-handling service. The product is called Banyan Intelligent Messaging. It is designed primarily to work with their network operating system.

DEC

DEC entered the workgroup computing market in early 1992 with Teamlinks, which allows DEC to deliver its Network Application Services

(NAS) to a Windows or Macintosh environment through the All-in-1 integrated office system.

IBM

IBM has entered the workgroup computing fray by developing LAN-based interfaces to its ImagePlus 400 document management system.

Lotus Corporation

Lotus Notes is probably the best known groupware package on the market, and, in actuality, the product only has a small handful of competitors. Lotus positions Notes as an effective, easy-to-use back-end system that provides the communication infrastructure to front-end information sources. In addition, Lotus has recently joined with Kodak to provide a document-imaging capability.

Microsoft Corporation

Microsoft's entry into the groupware market is Windows for Workgroups. Microsoft is hoping to build on the popularity of its Windows operating system by promoting its open-architecture development environment

WordPerfect Corporation

WordPerfect Office 4.0 is an integrated e-mail, calendar, scheduling, task management, and work-flow application. WordPerfect Corporation hopes to use its established word-processing base as a springboard into the groupware market.

Items to Consider

There are several items you must consider before deciding to deploy groupware throughout your enterprise. General words of wisdom: tread lightly and remain flexible.

Infrastructure

The first item you should consider is whether your infrastructure can support groupware applications. Network dynamics will change, storage requirements will change, and traffic patterns will be altered. An inadequate infrastructure could doom your efforts even before you begin. Note that you must consider both the LAN architecture and the computing environment.

Standards

Standards are probably the one big hindrance in the development of groupware applications. The industry is currently promoting several conflicting standards, primarily because vendors don't want to relinquish their proprietary edge. Make sure to ask vendors what they are doing to promote interpretability.

Culture

Organizational dynamics will change once people get accustomed to using groupware technology. The questions you have to ask are: Can the people adapt? Can management adapt? What processes will change? How will processes change?.

Where Is Groupware Going?

Groupware is targeted at the most precious resource in today's fast-moving organization—the workgroup. As the technology evolves, it will allow organizations to reduce overall cycle time, improve decisions, and empower knowledge teams at all levels. The technology itself, however, still has several major hurdles to overcome. In addition, most organizations will need to address overall process and work-flow design for any groupware product to become successful.

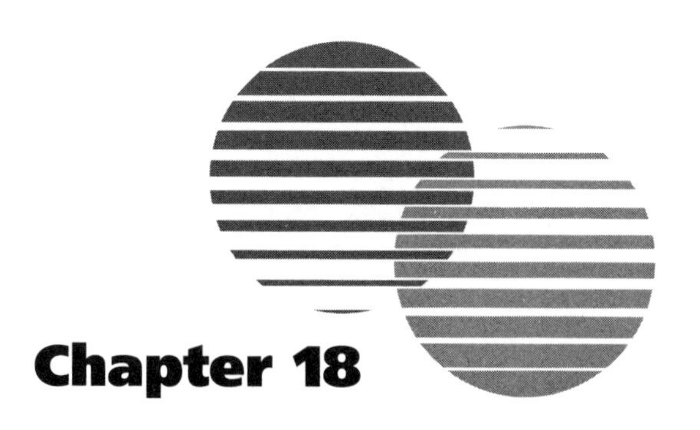

Chapter 18

Application Development Technology and Techniques

Application development has long been the private turf of the IS department. It defines the very essence of the department. In the glory days of IS, a mainframe in the glass house and a few long-term development projects on the schedule meant big budgets and job security to dozens of developers and supervisors. But that philosophy is changing now, along with everything else about IS.

More than ever, the IS department is under pressure to produce useful products and contribute to the betterment of the whole organization. Life cycles for mission-critical applications are growing shorter, meaning that IS needs to speed up the development cycle to get updated and accurate applications out the door quicker.

Development managers are turning to new tools, technologies, and techniques to produce better applications in a shorter amount of time. OOP, CASE, RAD, JAD: These are more than just funny acronyms; they are a new way of life for the application developer.

This chapter explores some of these new technologies and techniques. We look at ways to get the end user more involved in development and ways to get the computer to do most of the design and code generation. And, we touch on the new paradigm in application development, where everything is an object, rather than a discrete process.

RAD: Rapid Application Development

In 1989, James Martin, leading database expert and chairman of James Martin Associates, coined the term RAD: Rapid Application Development. "RAD is based on the use of Joint Application Design workshops, integrated computer-aided software engineering tools, small teams of highly trained and motivated users and systems professionals, a development methodology that defines the steps required to achieve high-speed development, and management techniques aimed at cutting through bureaucratic obstacles." In short, the components of RAD include the methodology, people, project management skills, and tools. Interestingly, Martin feels that 80 percent of RAD's effectiveness comes from better management processes, and only 20 percent is due to better computer technology.

The RAD life cycle has four phases: requirements planning, user design, construction, and cutover. The design specifications evolve from joint discussions, which we will discuss later. The result of these discussions is a detailed set of user requirements and application specifications that is used to develop a "production prototype" of the application. In this construction phase, the prototype is continually refined until it becomes the actual production application.

Cutover, or implementation, issues are addressed early in the development cycle. Continuous user involvement helps to plan for conversion, integration testing, documentation, and user training, speeding up the delivery of the final system. In traditional application development, these issues are not addressed until the application is ready to go into production mode.

RAD embraces the notion of "successive approximation"—that is, code a little, test a little—until the prototype becomes the final product. The result is (or should be) an application that closely meets the needs of the user community, developed in a relatively short period of time.

JAD: Joint Application Design

At the heart of RAD is JAD—Joint Application Design. JAD is a relatively new software development technique that addresses a common problem in applications development: the lack of effective communication between end users and IS professionals. Analysts and key end users participate in structured workshops to design specifications for computer applications. With the joint preparation of design specs, system development can proceed much more rapidly than it would under normal design procedures.

The process begins with discussion among the participants. The IS analysts translate users' requirements into relevant data models, screen and report designs, process flow diagrams, and decomposition diagrams. Specifications are captured by a scribe using an integrated computer-aided software engineering (I-CASE) tool, which checks specifications against a corporate model in a repository and prepares prototype applications for review. With each successive rendition of a prototype, the application gets closer to the final look. What's more, a good I-CASE tool will help to produce structured program code and documentation, relieving the IS programmers of much of the traditional development work.

Elements for Success with JAD

There are a lot of critical components that go into making a JAD session successful. First of all, the proper mix of end users is critical. They should represent a good balance between knowledge about the business and the authority to make decisions about the design, and they must be able to communicate well. In many instances, it is helpful for the participants to attend a class to learn how to participate in and contribute to the JAD workshop.

The JAD session leader is also very important. This person often must undergo specific training to learn how to properly manage a JAD session. It is his responsibility to keep the workshop on track and to lead the group to consensus about the application. Yet, he must be impartial and fair in dealing with any disagreements that may arise.

The I-CASE tools selected for the prototyping are also quite important. The tools must be flexible enough to incorporate both front-end diagramming facilities and a tightly integrated back-end code generator. With the proper I-CASE tool, the design developed in a JAD workshop can be converted quickly into a running prototype application that can be demonstrated to users. Some of the more popular tools in this category are Application Development Workbench (ADW) from KnowledgeWare Inc., Information Engineering Facility (IEF) from Texas Instruments Inc., and Information Engineering Workbench (IEW) from Bachman Information Systems Inc.

The IS analyst designated to use the I-CASE tools should be well versed in database design and normalized data models. He should also be skillful with the tools and techniques used to build and edit the programs, create the design, extract repository information, build screen designs and reports, and create prototypes.

Finally, the atmosphere of the workshop is quite important to success. The room should be equipped with overhead projectors, flip charts, copying facilities, the computer with the CASE tools, and a large-screen monitor. The seating arrangement should be U-shaped for maximum interaction among participants.

The JAD design process can be much faster than traditional IS analysis techniques. Moreover, users get more involved in developing the specifications for their applications, reducing cumbersome review cycles during development. JAD harnesses the knowledge of end users and IS professionals, and it cuts across organizational boundaries. Furthermore, the I-CASE tools enforce structure and ensure quality. The end result is an application that closely meets users' needs, developed in a short time.

A JAD Case Study: An Investment Firm

Fidelity Systems Co., a subsidiary of Fidelity Investments of Boston, MA, is a believer in the JAD approach to application development. The firm's complex distributed computing environment incorporates six IBM-compatible mainframes, six AS/400s, three System/38s, 23 VAX systems, 11 Stratus machines, and more than 4,000 desktop devices. With such a diverse environment, the company wanted to speed product completion while maintaining data integrity among multiple databases on the various hosts.

Fidelity's new method for application development, built on the principals of JAD, rapid prototyping, and enterprise-wide data modeling, is called Fidelity Advanced Systems Environment (FASE) 2000. George Hathaway, vice president of software development at Fidelity Systems, says the company is looking to cut the time to deliver new systems in half by 1995 and to improve the quality at the same time.

The FASE 2000 method is platform independent, and so it can be used to create mainframe, minicomputer, or PC applications. A common data model ensures data consistency across systems. The first tangible product to come out of FASE 2000 is a client/server accounting application, which took just six months to develop. By comparison, Fidelity's old standard development cycles stretched out for 12 to 18 months.

A Case for CASE

While RAD and JAD are really methods or approaches for developing applications, computer-aided software engineering (CASE) is a tool for analyzing, designing, and building applications. Many traditional CASE tools are based on the paradigm of mainframe-based computing, but more and more tools for client/server applications are coming to market. These tools do everything from building graphical user interface screens to generating the underlying communications code between clients and application servers. This bevy of client/server-oriented CASE tools has appeared in response to corporate IS shops that are demanding an easy way to launch their own forays into client/server computing.

Roger Berry, vice president of IS at Houston, Texas-based Tenneco Gas, believes that CASE tools are beginning to have a profound impact on the role and responsibilities of IS programmers. Since the tools are now doing much of the design and code generation work, the programmer is relieved of some of the drudge work of development. He can now devote more of his time to business analysis, plugging IS into the business processes.

OOP: Object-Oriented Programming

Perhaps one of the hottest new programming approaches in recent years is object-oriented programming, or OOP. It is a new concept in which events—such as opening a file or clicking a mouse button—and the data associated with those events are treated as objects. Such objects can be coded as modules that can be reused. Developers design with and build upon these reusable code modules, which encapsulate a specified function through a clearly defined object interface.

A basic premise of OOP is to create these modules once and use them again and again in multiple applications. Common sense tells you that this approach will greatly increase development productivity—once the programmer has a sufficient library of object code. That's one of the hitches in OOP development. Programmer productivity can be extremely low in the beginning as the developer learns the object paradigm and creates (or purchases) his object libraries. Once the programmer is over that hump, productivity tends to skyrocket.

Scott Koehler, a principal at Koehler Consulting in Holliston, MA, says the migration from procedural development to object technology is not really a language or syntax issue: "It is a fundamental change in the way software is developed. To exact that change in the minds of experienced software developers requires a radical break from past methods." In fact, it can be quite difficult for an experienced procedural code developer to make the transition to OOP. The change in fundamental concepts is just too great to overcome.

An object-oriented program is decidedly non-procedural. Objects and events can relate to each other in seemingly random ways. Moreover, an application isn't one monolithic block of code; it's hundreds of discrete objects. By comparison, a procedural program is a block of code with a definite execution routine. First *this* happens, then *that*. The order is always the same.

The Pitfalls and Promises of OOP

The biggest hurdle for any company that is getting into OOP techniques is training programmers. As we have already said, the problem lies not just in learning a new language or new syntax, but in grasping new concepts.

A second problem with OOP is the current dearth of off-the-shelf business-oriented class libraries—existing code that describes objects and their capabilities. As more class libraries become available, corporate developers can spend more time assembling applications from prebuilt classes and less time writing, testing, and debugging new classes of objects.

The good news is that there are plenty of tools available. There are editors for revising code, debuggers for tracing and correcting errors, and browsers for analyzing the structure of the code and the relationships of the objects. The tools, however, are far from mature. Many tools cannot be used together easily, and many offer limited platform support and rudimentary features. Moreover, standards for tools and class libraries are lagging.

Commercial software developers have made great strides with OOP, but most corporate IS shops have not. This situation is attributed chiefly to the heavy investment in training that is necessary to become proficient with the tools and techniques. Corporate IS is cautious in its spending. The potential benefits, however, are good enough to suggest that IS shops should be at least dabbling in OOP development today.

And just what are those benefits? For one thing, just like anything else, it comes down to money. It stands to reason that if applications can ultimately be developed with reusable code in a shorter period of time, the organization can save money. In addition, the modularity of the applications simplifies design. It makes finding errors easier and reduces maintenance requirements.

Going Graphical

Graphical user interfaces (GUIs) on desktop PCs and workstations have changed the way that people interact with computers. James Martin says that GUIs are extremely important for new application development: "Study after study has shown that end users learn new applications more quickly and make fewer mistakes if a GUI is well designed." This move to GUI environments goes hand in hand with the trend toward client/server computing.

The shift to graphical environments has created new application development requirements for both IS professionals and software tools. Vendors of development tools are coming out with new or revamped products and strategies. While current CASE tools are mostly still behind the curve in this move to GUI development, there are a number of tools that are specifically designed for building GUIs. Some work in conjunction with traditional CASE tools, while others work independently. The most productive products generate the program code as well as the user interface. Representative GUI builders are UIM/X from Visual Edge Software Ltd., Builder Xcessory from Integrated Computing Solutions Inc., AIX/Interface Composer from IBM, and Tigre Programming Environment from Tigre Object Systems Inc. ShortCut, marketed by Cadre Technologies, Inc., is designed specifically for generating prototypes of GUIs in JAD sessions.

Mikael Wipperfeld is director of product marketing for Informix Software, Inc.'s environment business unit. Wipperfeld has identified five criteria for development tools and CASE products in a GUI-based computing environment:

1. The tools must allow users to migrate from their existing legacy systems and applications to the new platform and environment.
2. The tools should generate source code that is independent of a specific user interface or GUI.
3. GUI programming tools should aid in building applications for specific industries.
4. The programming languages should generate prototypes that will in turn generate almost complete programs.
5. The tools should aid end-user access to databases.

Xbase Marks the Spot

The Xbase language is alive and well for the Intel-based workstation market. Xbase is a generic language derived from Borland International's dBASE programming language. There are a number of products and tools in the Xbase domain, including compilers, report generators, code libraries, editors, debuggers, code generators, and full-blown integrated relational database management systems.

Xbase is a popular programming language for small to medium-sized applications. Few companies develop enterprise applications with this language. Yet, for personal or workgroup applications, Xbase satisfies most needs. Many of the Xbase products include application generators—integrated modules that write the code for complete applications. These easy-to-use application generators make it possible for non-programmers to develop their own applications quickly on a foundation of standard, structured code. Some of the more popular Xbase language products that include code generators include dBASE IV from Borland International, FoxPro from Microsoft Corporation, Genifer from Bytel Corporation, RAD from Hilton Productions, and Scrimage from Synergy.

COBOL: It's Not Dead Yet

Despite proclamations from many computer industry pundits, the COBOL language is not dead yet. It seems as if everyone from Phillipe Kahn to your Aunt Minerva has been predicting the demise of COBOL development, as IS shops flock to newer technologies and languages. But COBOL lives on, not necessarily for new development, but certainly to maintain and support legacy systems. There are over 60 *billion* lines of COBOL code out there, folks, and they're not going to disappear overnight.

The downsizing trend has given rise to a rash of "new" COBOL applications on PCs and networks. Products like Realia's MicroFocus COBOL help companies port existing applications nearly intact from one platform to another. Many companies use this approach as a stopgap measure until they can re-engineer and rewrite the applications.

Tenneco's Berry says his shop still uses COBOL on occasion, particularly when the application logic is too convoluted for development in other languages such as C. Berry says, however, that less than five percent of his department's new development is in COBOL.

The Real Issue Is Integration

For most IS shops, the real application development issue is integration. Every IS shop has become or is becoming a systems integrator. New systems need to be blended with legacy systems. Departmental applications need to tie into enterprise applications. Desktop and network platforms are merging with mainframe and minicomputer platforms. The challenge is to make it all work together without starting from scratch.

There are development tools on the market that assume that most new systems must integrate the old with the new. The old systems consist of the legacy database technology investments. The new systems include graphical user interfaces and client/server technology. The marriage of the old and the new takes place through APIs—application programming interfaces. Two prominent products in this category are Microsoft's Visual Basic and Borland International's ObjectVision. Both tools provide for designing powerful visual front ends to existing databases on a variety of hardware platforms.

The Face of Development Is Changing

Application development is changing in the attempt to keep pace with business changes: Prototyping for rapid application development. Computers aiding in the development of applications. New development paradigms based on the encapsulation of data and activities. These new technologies and techniques are aimed at getting better applications into the hands of the users at a faster pace.

Chapter 19

Client/Server Computing

Client/server computing represents the latest evolution in the use cycle of the desktop computer. In the early to mid 1980s, desktop computers were primarily used in a standalone mode. They were great for processing their own applications, but they didn't communicate with any other computers. Then along came the 3270 adapter card, and voilà, the PC became an expensive dumb terminal with access to the mainframe. Oddly enough, we essentially disabled the PC's processing capabilities with this configuration.

The next step forward came with local area networking, which linked standalone PCs together for the purpose of sharing files and peripherals. The individual PCs did all the processing, while shared file servers stored applications and user files. While this represented a great improvement over standalone computing, it still did not offer a solid computing platform for enterprise applications. There were lingering issues such as data security, network traffic, file contention, and the like.

Finally, client/server computing arrived to usher in the era of enterprise network computing. At last we have found a way to exploit the power of our desktop PCs fully while having all of the advantages of "big system" computing. Client/server computing separates the functions of an application into two distinct parts. The "front-end" client component presents and manipulates data on the desktop workstation, using familiar PC tools. The "back-end" server component acts like a mainframe to store, retrieve, and protect data. The two ends share a true division of labor, with parts of an application executing on different systems.

In this chapter, we look at how client/server computing works and how it differs from other types of computing. We discuss the advantages as well as the caveats and lessons from the front lines of the bleeding edge. We also explore how client/server computing can change work roles for people, and we present Seven Suggestions for Successful Client/Server Computing.

How Client/Server Computing Works

We have already described client/server computing as a division of labor. Parts of an application get processed by different computers, taking full advantage of the intelligence of desktop computers as well as the power of the central "back end" computer.

The "client" part of the equation is usually a desktop computer such as an IBM-compatible PC or an Apple Macintosh. UNIX workstations are also commonly used as clients. It is essential that the desktop workstation have processing capabilities of its own; a terminal is not considered to be a client because it has little or no "intelligence."

The "server" part of the configuration is another computer that is largely defined by its role and responsibilities. The server can be anything from another PC on up to a mainframe computer. In client/server processing, clients are attached to servers and request application-specific actions or services from the server.

In his article "The Client/Server Paradigm: Making Sense out of the Claims," author James Buzzard says there are a few attributes that further define client/server computing. True client/server processing requires:

1. Communication between the client and the server
2. Client-initiated interactions with the server
3. Restriction by the server over the client's authority to request services or data from the server
4. Arbitration by the server over conflicting requests from multiple clients
5. Division of application processing between the client and the server

Buzzard claims that most file server-based systems meet the first four criteria. The fifth point is key in determining true client/server functionality: the application's logic is divided so that part is executed by the client computer and part by the server computer. Database applications are the most common type of client/server applications. Electronic mail and document management are other common uses.

Comparing Systems

Let's look at how a database application is treated on a client/server platform, by contrast with more traditional platforms. We'll use the example of accessing an online telephone directory to get a listing of all the people named John Smith in the city of Houston.

In a traditional mainframe configuration, all processing power is centralized. When a user sitting at a terminal (or a PC configured as a terminal) queries the database for the listing of all the John Smiths, the mainframe accesses its data repository and comes up with the listing of just the John Smiths. The mainframe then sends a *screen image* of the answer down the wire to the waiting terminal. What's important here is that *no data* are actually sent to the terminal, only a picture of the data. The terminal cannot do anything with the data except display them, as you can see in Figure 10.

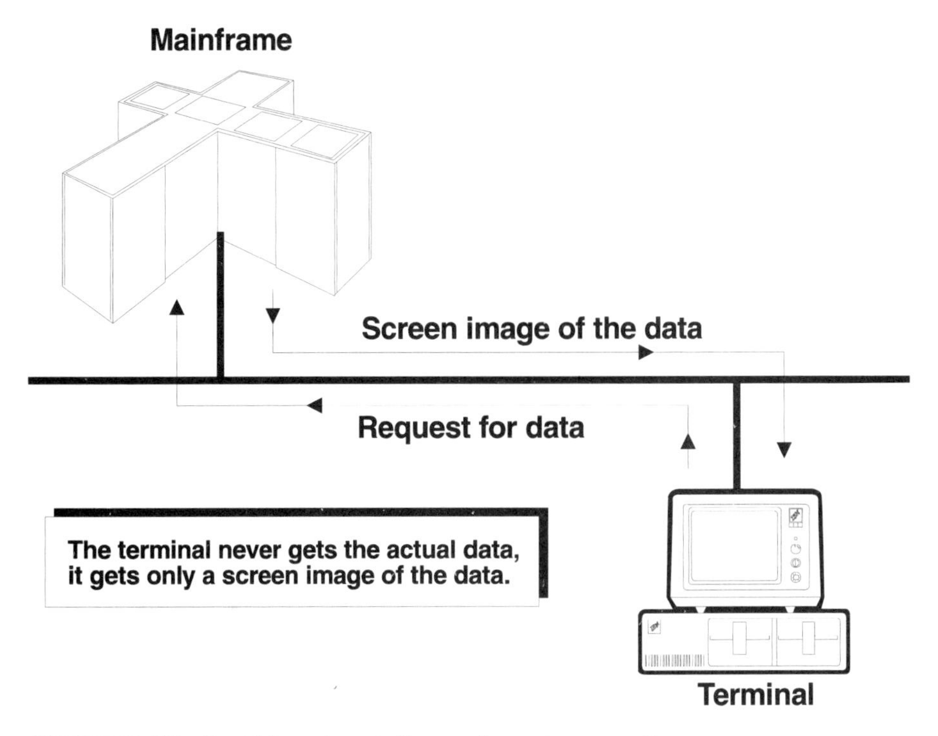

FIGURE 10 Traditional mainframe-based computing

In a traditional local area network file server configuration, the processing shifts to the client. When a user sitting at a PC queries the database on the file server for the list of John Smiths, the server sends the *entire phone directory* down to the client. The client then uses its processing brains to look up what it wants and comes up with just the John Smiths. The key here is that the server is not doing any data screening; it is simply storing the data until the client workstation asks for them. As a result, very large files are transmitted over the network wire when a client needs data. See Figure 11.

In our client/server configuration, the client initiates its request for the list of John Smiths. The database server takes that request, searches its data banks, and comes up with the very specific list of John Smiths. The answer is then transmitted down the network wire to the client. The key points are that the server has done the processing to send actual data to the client. What's more, only a very small file (the answer to the request) is transmitted

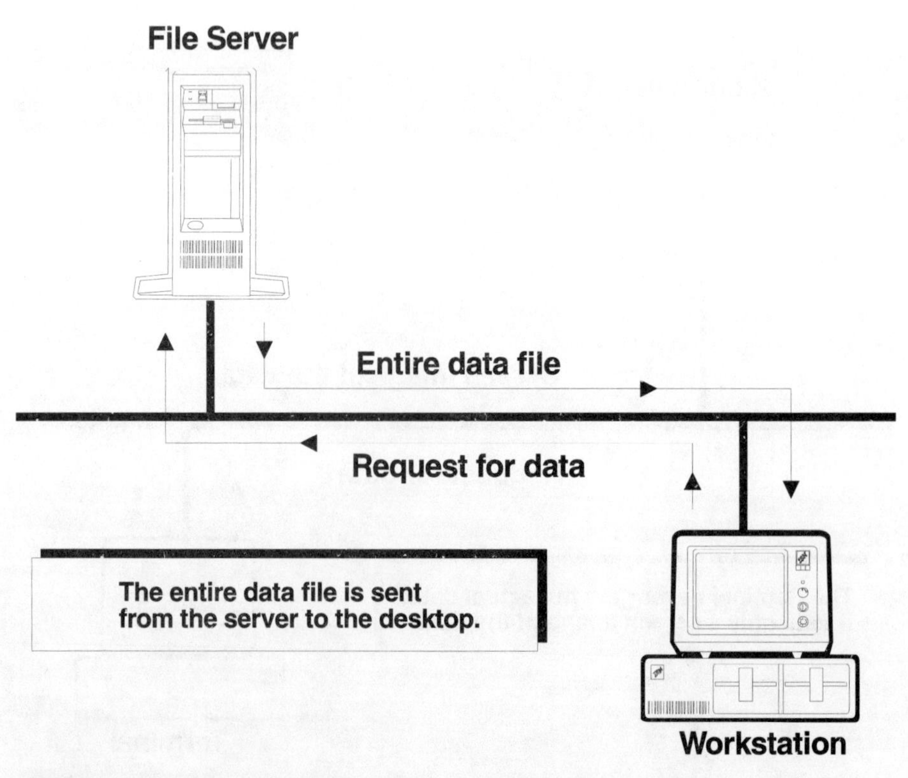

FIGURE 11 File server-based network computing

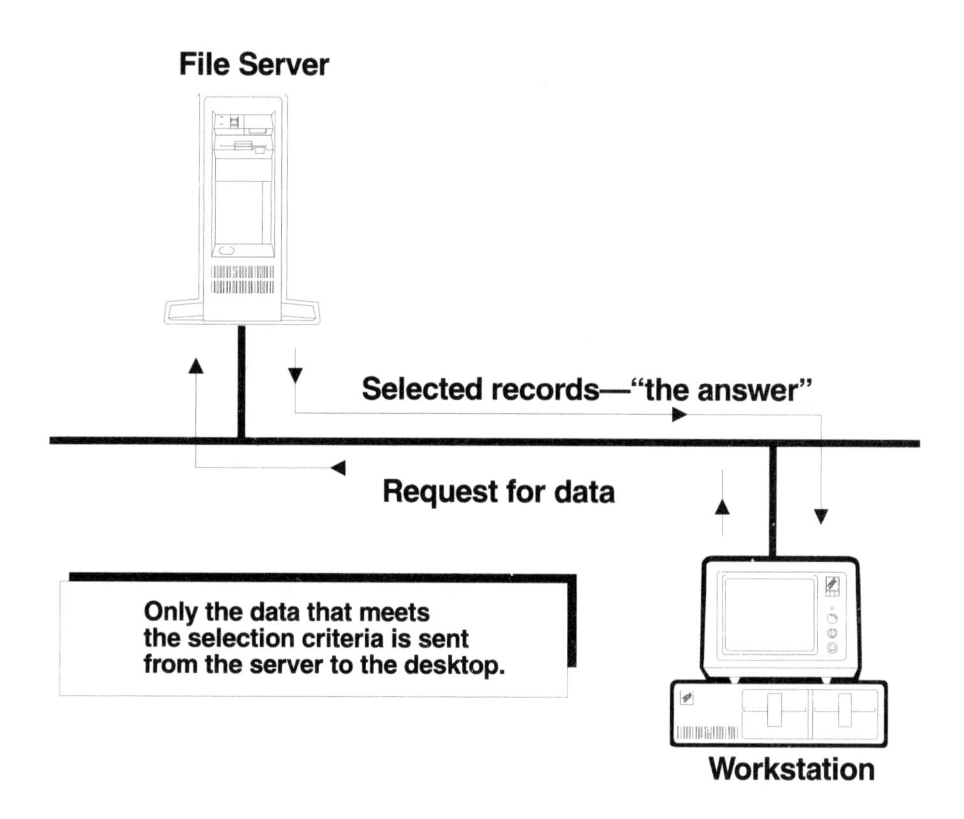

FIGURE 12 Client/server-based network computing

via the network. The client can now use the data for reporting, for further manipulation, or for any other purpose. See Figure 12.

A client/server configuration usually includes a special-purpose server called a database server. The database server is usually a separate machine from the one that runs the network operating system. The purpose of the database server is to store, protect, and manage data.

Client/server computing keeps the task of processing close to the source of the data being processed. In the case of a database application, the data processing is done on the computer that holds the data (the database server). The user interface processing (e.g., the presentation of the data) is done on the desktop computer that directly interacts with the user. This localization of processing can greatly reduce network traffic, increasing effective throughput and carrying capacity on heavily used networks.

The Advantages of Client/Server Computing

Client/server computing has several characteristics that give it a distinct advantage over other computing choices:

- The overall hardware, software, and maintenance costs are lower than for mainframe or minicomputer systems.
- The strategy maximizes the use of intelligent workstations to divide the labor of the application, lessening the requirements for large backroom computers.
- There is centralized data control at the database server level, providing for a secure environment with integrity checks and balances.
- Data elements and program code are reusable from application to application, reducing development time and cost.
- Data transfer over the network medium is limited, minimizing response time and optimizing network resources.
- The systems are scalable and relatively open, allowing for flexibility in a rapidly changing environment.

Who Is Using Client/Server Computing?

Most Fortune companies are at least dabbling in client/server development, and many are actively involved in it. The move to a client/server platform often goes hand in hand with downsizing and reengineering efforts. (See Chapters 9 and 12.) We could give countless examples of how companies have implemented client/server computing, but we'll limit our case studies to just a few.

Client/Server Case Study: A Utility Company

Northern Illinois Gas is implementing a client/server application in order to improve its natural-gas distribution system. The company is moving its gas flow computing operation to a PC-based LAN. The network will run a custom-built graphical modeling application and store the models on a fault-tolerant storage device. Ultimately, the application will permit

Northern Illinois Gas to make rapid decisions regarding the flow of gas to customers.

The current method for analyzing data for these decisions is cumbersome and time consuming. Analysts pore over stacks of printed gas-load data generated by the company's mainframes. The data are used by gas distribution system engineers to draw maps and make judgments about installing new lines and maintaining and upgrading existing lines. Unfortunately, the calculations take too long and are too labor intensive for the engineers to evaluate all possible scenarios. Moreover, the hand-drawn maps get outdated quickly.

The new system will give desktop access to graphical models of the pipeline system derived from data on the mainframe. Various scenarios can be analyzed in as little as half an hour, compared to days under the old system.

The utility company is using custom-built client/server software for the application. Data are extracted from the mainframe, and the PC software converts the data into working models and maps of the distribution system, making the information much easier to decipher.

Northern Illinois Gas has this client/server application in test mode, with full implementation expected by the third quarter of 1993. The $1.4 million system is expected to pay for itself within just two years. The company anticipates various benefits from the system, including better analytical abilities leading to cost savings, and the ability to respond quicker to new customers' request for service.

Client/Server Case Study: A Railroad

Burlington Northern Railroad recently implemented a client/server application that allows railroad operations specialists to schedule trains and optimize their movement over 25,000 miles of track and through 8,000 terminals and stations across the country. The new system extracts data from a mainframe and downloads it to an IBM RS/6000, where LAN-based PCs can initiate requests for data for inclusion in charts, documents, or spreadsheets.

One of the reasons the railroad decided to implement such a system is the painfully slow response time of the mainframe. Some interactive requests took five to ten minutes or more on the mainframe. The same requests can be processed by the RS/6000 and delivered to the users' desktops in 15 to 30 seconds. The company also likes the graphical capabilities and the friendly user interface of the client/server system. Moreover, one planning specialist estimates that the new environment has boosted development productivity by 85 to 100 percent.

Selecting Tools: The Leading Players

Almost every software product on the market today makes some sort of claim that it is designed for client/server computing. Selecting the appropriate tools for your environment and applications can be tricky. It is particularly important to select front ends and back ends that work together. We recommend selecting one or two good back-end database products and then selecting the front-end tools to work with them. Table 11 highlights some of the most popular back-end database products on the market today.

TABLE 11 Database Back-End Products

Product	Vendor
Oracle Server	Oracle Corporation
Interbase	Borland International
Ingres Intelligent Database	Ingres Corporation
Informix Online	Informix Software
SQL Server	Microsoft Corporation/Sybase, Inc.
SQL Base	Gupta Technologies
NetWare SQL	Novell Inc.
XDB-Server	XDB Systems
Progress	Progress Software
PROBE	Prologic Computer Corporation

There is a myriad of products that can be used on the front end, or client. Some are development tools designed for IS professionals; others are productivity tools with imbedded structured query language (SQL) code, designed for ease of use by end users. Table 12 is a partial list of popular front-end tools. Space does not permit us to list all products on the market today.

TABLE 12 Database Front-End Products

Product	Vendor	Intended User
Lotus 1-2-3	Lotus Development Corp.	End User
Excel	Microsoft Corp.	End User
Paradox	Borland International	End User, Developer
Oracle Card	Oracle Corp.	End User
Quest	Gupta Technologies	End User
ObjectVision	Borland International	End User
Q + E	Pioneer Systems	End User
Forest & Trees	Channel Computing	End User, Developer
Visual Basic	Microsoft Corp.	Developer
Access	Microsoft Corp.	End User, Developer
FoxPro	Microsoft Corp.	Developer
PowerBuilder	Powersoft Corp.	Developer
Oracle SQL Forms	Oracle Corp.	Developer
SQL Windows	Gupta Technologies	Developer
dBASE Server Edition	Borland International	Developer
Approach	Lotus Development Corp.	End User
Focus	Information Builders	Developer

We suggest that you test and select a range of front-end products, keeping in mind that you will need tools for end users, programmers, and developers. Always keep an eye on the software market, as new client/server products are coming to market weekly.

There is also a growing market for middleware products—products that bridge the gap between front ends and back ends. Middleware products promote industry standards so that developers can write applications to the standard specifications, without regard for which front ends and back ends may be in use. Thus, they help to increase the portability and compatibility of applications.

Changing Roles and Updating Skills

Client/server computing will usher in more than new technology. It will also bring new roles for IS staffers and end users, which in turn will require new skills. In making the transition from a traditional environment based on central processing to the distributed client/server environment, the collective IS staff will need to adopt new programming techniques, learn more about networks and communications in general, and off-load report writing tasks to users.

IS roles will change—from data guardian to data expediter. End user roles will change as well—from data recipient to data owner. The goal will be to get usable information into the hands of the people who need it, when they need it.

Gary Senese, vice president of MIS at Rubbermaid Commercial Products, Inc., describes the process of changing IS as "morphing." That is, programmer/analysts will need to change into business analysts. Senese says that not changing is not an option, for change is "the only way we can continue to add value to our organizations commensurate with our costs."

First Retrain...

When client/server computing is fully implemented, the mix of skills required in the average IS department will change. Existing staff will have to be retrained or replaced. The prudent manager will set aside funds and time to bring the staff up to speed on PCs, LANs, PC applications, network and systems management, application development, prototyping, and support for end-user applications. Former mainframers will need to learn how to handle asset management, configuration management, performance analysis, archiving, backup, and other management activities in network-based environments.

The training, while necessary, won't be cheap. According to Forrester Research, the average cost to retrain an IS staffer is $12,000 to $15,000, with an annual continuing education cost of $1,500 to $2,500 per person. Is the investment worth it? Many companies think the alternative—doing nothing—is much more expensive in the long run than paying for the change now.

End users, too, will need training to learn new data access tools. They may not necessarily need to learn SQL code, but they should know how to assemble the data they need. Companies also may find it necessary to offer "data" training. End users have been accustomed to having limited access to data or to looking at it in only one format. Now they need to know more about what data are really there, so that they can decide for themselves what they want and how they want it.

End-user training can really pay off in increased productivity. Studies have shown that users who control and format their own information and determine their own access requirements are at least 50 percent more productive and about 70 percent more satisfied with their systems.

...Then Reorganize

John McCarthy, director of computing strategy at Forrester Research of Cambridge, MA, suggests that companies moving heavily into client/server computing should reorganize IS resources. McCarthy advocates having one internal group to support individual business units and another to maintain corporate standards. Legacy systems, says McCarthy, should be outsourced. This reorganization will allow companies to redistribute budgets, allowing their purchasing departments to buy commodity-level computer equipment, while business units become responsible for the cost of developing applications for their individual needs.

Lessons from the Front Line

By this time, many companies have entered the client/server fracas, and, though most of them report great benefits, they say it wasn't without a lot of blood, sweat, and tears. Here are a few of the caveats, concerns, and lessons from the front.

- In-house programmers have to learn new products, new tools, and, sometimes, whole new ways of approaching a development task. They must learn to work in teams and do more planning and analysis up front. The transition doesn't happen overnight.
- It can be difficult to find qualified help for designing and developing the applications and for optimizing the network. Scrutinize outside help closely before hiring consultants.
- Maintaining decentralized, distributed applications can be a real challenge. The developers must formulate a plan for distributing software updates to remote servers and clients.
- Network strategies may need to be updated to be able to send the least amount of data between server and client in order to minimize network traffic.

- The four most common operating system platforms (DOS, Microsoft Windows, OS/2, and UNIX) each have drawbacks that make application development tricky. Future enhancements, as well as new operating systems like Windows NT and NextStep, may stabilize development and make the applications easier to use.

Computerland's Seven Suggestions for Successful Client/Server Computing

The January/February 1992 issue of *Computerland Magazine* lists seven basic issues that should be addressed for a smooth transition into client/server computing:

1. *Planning:* Applications should be carefully planned before any development begins. It's best to use a task force of IS and end-user personnel to collect application requirements, set standards, and institute pilot programs.
2. *Compatibility:* When selecting front-end and back-end tools, be sure to choose compatible products.
3. *Rollout strategy:* Introduce the new architecture in pieces or by workgroups.
4. *Support:* Network support functions should be centralized if possible. Full-time network professionals should perform the work.
5. *Budgeting:* Expect to spend money on people and specialized outside services. Cost savings on hardware may have to be rechanneled to software and personnel expenses.
6. *Gains:* Productivity gains won't happen overnight. Remember that IS personnel and end users alike will need time to adjust to the new environment.
7. *Jobs redefined:* That end users access more data means workers can do things differently and can do different things.

Client/server computing represents the next logical step in distributed data processing. It takes advantage of the computing resources on the desktop as well as in the back room. The advantages loom large: potential cost savings, increased flexibility, better use of the resources, and more. It is easy to see why so many organizations are making the transition to client/server computing.

Chapter 20

Consultants: Using Outside Help Effectively

Throughout this book, we have talked about change and keeping up with new technologies, new strategies, and new challenges. Needless to say, that can be a pretty intimidating task. Fortunately, you don't have to go it alone. There is a large industry built around people who are there to help you when help is needed. Whether you call them consultants, advisers, outsourcers, or contractors, they bring to the table a mix of skills, experience, and knowledge that may be the edge in making your projects a success.

But simply hiring a consultant doesn't guarantee success. You need to know how to use that consultant effectively to meet some of your specific goals. This chapter is devoted to providing you with tips on how to select and hire outside help and how to use that person or group to your best advantage. We focus on four types of outside help: outsourcing agencies, contract programmers, management consultants, and technical consultants. For simplicity's sake, we refer to all four of these groups as "consultants." Moreover, our use of "consultant" can refer to a single person, a group of people, or an entire company.

Why Hire a Consultant?

Some companies rely heavily on consultants' advice and help, while other companies shun consultants. That is a matter of personal (or rather, corporate) preference. The fact of the matter is that everyone could benefit from hiring a consultant at least once in a while.

Perhaps the most common reason for hiring a consultant is to obtain that person's or group's special knowledge or skill for a period of time. Using a consultant gives you the opportunity to get what you need, when you need it, without having to worry about extending full-time employment. Consultants may be hired to temporarily fill open work slots, perform drudge work, or meet the demand of excessive work loads. Or, more likely, they can perform a highly specialized task or service that you cannot carry out with in-house resources.

People who are outside of your own company or organization have a unique virtue: they aren't bound by your corporate politics. This fact gives consultants an objectivity that you can't find in house. The consultant can cut through the BS and not worry about CLMs—career limiting moves. We call this B-52 consulting—being able to drop the bomb and walk away unscathed. Sometimes a consultant's "bomb" is just what is needed to get your organization off dead center and onto progressive action.

A consultant can give you new perspectives on a subject. Your own people may be too close to a problem or situation to consider alternative plans of action. A consultant can scc things from different (and objective) angles and can make suggestions that you may not have thought of yourself. Likewise, you could hire a consultant to give you a second opinion on an action plan. As a consulting firm, Currid & Company is often called upon to review IS strategies and give our opinions and recommendations.

Technical consultants can be excellent sources of information. You can use outsiders to share knowledge and help to get your own staff up to speed on technical matters. This gives you the benefit of "cross pollination"—that is, sharing the benefit of the consultant's experience with other companies. Of course, don't expect the consultant to share sensitive information about other organizations. Most of us do have scruples!

Before You Hire

Before you hire any consultant, you should carefully assess your needs and understand what you hope to get from the consultant. Make sure there is a consensus in house as to what you want to gain. You certainly don't want different sides of the house giving mixed signals.

Regardless of the size of the project, you should write down your objectives for hiring an outsider. Document the deliverables you want to receive. The more information you have collected about your project, the easier it will be to find just the right outside person or firm. Keep in mind that different consultants have different areas of expertise. The more specific you are about your needs, the more closely you can match your requirements to the consultant's skills.

It has been Currid & Company's experience that many clients call us with a very vague notion of what they want from a consulting firm. We are often asked, "What do you do, and how can you help us?" All too often, the answer to those questions determines what the client defines as his requirements. While we usually enjoy the ensuing work, we do recommend that companies have a good idea of what they want before soliciting proposals.

If your project looks like it will involve quite a bit of work, you may want to write a formal Request for Proposal (RFP). This is standard practice for most government agencies. The RFP can then be sent to several consultants, and the resulting proposals can be weighed against each other.

The RFP should explain what you want done, as well as any details of how you want the work done. It should include very clear specifications and expectations about the job, including all final deliverables. It's also helpful to include as much background information as possible about your organization and the specific project. We also suggest you specify exactly what information you want people to provide in their proposal. Before you send the RFP to potential consultants, it is helpful to contact them to see if they are capable or even interested in performing the work. Thus, you save the time and expense of soliciting bids from inappropriate vendors.

Those consultants that are interested in performing your work should respond with a formal proposal. As a minimum, you should ask them to include these items in their proposal:

- A management summary that includes brief statements about the consultant's understanding of the project, his approach to doing the work, and his estimated cost
- A detailed explanation of the consultant's project implementation approach
- The resulting products to be delivered throughout or at the end of the project. Deliverables may be reports, presentations, application systems, or whatever else you are seeking

- A schedule for reaching milestones and completing the project
- The estimated cost for completing the project. Cost estimates should include professional fees and related business expenses.
- A manner of dealing with project scope changes, should you decide to shrink or expand the workload once the project is under way
- The consultant's qualifications for performing the work, including experience on similar projects
- References from previous clients

How the consultant responds to your RFP should be a clue as to how he might perform your work. If the proposal is complete and well organized, you can be fairly assured that he will take the same approach to your project. If the proposal is incomplete and poorly organized, you can infer that his work may be of similar quality.

As you evaluate each consultant's proposal, pay considerable attention to his implementation approach. Is it sound and reasonable? Is it an approach you would undertake if you had the resources in house? Is there innovation and creativity that would add value to the project? Will the consultant's approach meet your needs? Going beyond the approach, look at the consultant's experience with similar projects. Is he familiar with your technology? Does he understand your industry? The more knowledgeable the consultant is in these areas, the more likely you are to receive appropriate deliverables that meet your needs.

If you are going to solicit proposals from small consulting firms, you may want to consider paying a small honorarium for the proposal. The fact of the matter is that small firms often cannot afford to spend the time developing a proposal, even for a job for which the firm is well qualified. Unfortunately, cash flow is king, and spending days on a quality work plan and proposal can actually hurt the firm in terms of lost revenue opportunity. You may be missing out on a highly qualified consultant for this unfortunate economic reason. Likewise, if you are going to ask an out-of-town consultant to visit your location to make a presentation regarding his proposal, consider paying for his travel expenses. It's a small price to pay to find the most appropriate consultant for your job.

Using Consultants Effectively

Early in the chapter, we said we would use the generic term "consultant" to refer to several types of outside help. In the next few pages, we'll get more specific about each type—how to hire them, and how to use them.

Outsourcing Agencies

Chapter 10 addresses the topic of outsourcing—using an outside agency to perform specific tasks for a defined period of time. It is very common today to outsource all or part of IS operations, from application development to operation and support.

Tips on Hiring an Outsourcer

You definitely should use a well-defined RFP to solicit proposals and bids from outsourcing agencies. Your RFP should get very specific about what services the outsourcer is expected to provide. You should also detail your desired performance levels and how you plan to evaluate the contractor's performance. In fairness to the bidders, you should also point out any penalties for not meeting those performance levels.

If your company is like most organizations, you are outsourcing work in order to save money. Therefore, you should evaluate the outsourcers' proposals very closely in the "costs" area. Many companies grow disillusioned with their outsourcing agency because of escalating or unexpected costs. You are well advised to head that problem off before it starts by being aware of all projected costs before you sign a contract. And though you can't be expected to know, for example, what your transaction processing level might be five years from now, you can ask the outsourcer for a fee schedule for increased transaction levels.

The Most Effective Use of Outsourcing Agencies

An outsourcing agency is best used to perform non-strategic, routine work that can be easily defined and measured. Companies commonly hire outsourcers for application development; computer operations; help desk support; training; network and workstation installation, operation, and maintenance; communications; data entry; and other non-strategic services such as word processing, record keeping, publications, and documentation.

You *don't* want to use an outsourcer for strategic applications or tasks, or for setting your IS strategies. Remember that information technology can be a valuable competitive weapon, and you don't want to give an outsider control of your most important weapons.

Special Considerations When Hiring an Outsourcer

Companies usually enter into long-term contracts with outsourcing agencies, sometimes lasting five to ten years. That's a long time to be allied with a specific business partner. Make sure you have a detailed contract that specifies all the work you expect the outsourcer to perform, as well as how you intend to measure his performance. Establish specific service levels, the penalties for not meeting those service levels, and any bonuses or incentives for exceeding the service levels. Also, allow yourself an "out" if the partnership doesn't work out. Try to avoid excessive penalty fees for terminating the contract prematurely.

Establish a workable reporting structure that doesn't create too much of a bureaucratic obstacle. Keep the channels of communication open to avoid an adversarial relationship. And, be careful of having too many people give direction to the outsourcers. That's a prime recipe for losing control of the resources and ending up with cost overruns.

Contract Programmers

Contract programmers differ from outsourcing agencies in that they usually are independent or represented by a development "body shop" that places individuals for a specific programming assignment. The skill of the individual is key in her getting the job.

Tips on Hiring a Contract Programmer

An RFP is not necessary in hiring an individual contract programmer. However, you should ask to see examples of her previous work. If you can't view applications themselves, then ask to see samples of her documentation or a demo disk of various applications. A good developer creates complete, organized documentation that anyone can understand. The demo disk will give you an idea of her programming style. Do her applications use function keys? Icons? On-screen help? Also look at her experience level with languages, products, and tools that you plan to use for your application. You are hiring this person for her expertise, not training her for her next assignment.

The Most Effective Use of Contract Programmers

Contract programmers can be used to fill the gaps that your own internal resources can't. Use contractors for "quick hit" situations as opposed to long-term projects, unless you are using them to deal with a budget exigency. (Some companies find they can capitalize work done by contractors, but not work done by in-house personnel. This has an impact on who can do the work.)

Tenneco Gas in Houston, TX, is taking an innovative approach to using contract-programming labor. Roger Berry, vice president of IS, has assigned all support and maintenance of his mainframe-based legacy systems to contract personnel. For his new development on a client/server platform, he is using his own in-house staff. The reason? Berry wants to make sure that his own people get trained on the new C/S tools and techniques, while non-permanent programmers tend to the mundane old applications. Eventually, these legacy systems will be rewritten for the C/S platform, and the in-house staff will be ready to perform the task. The contract programmers can be let go without a serious loss of skills and knowledge.

Special Considerations When Hiring a Contract Programmer

Establish up front who owns the source code that the programmer develops. If you agree to use compiled or object code from the programmer, make sure to negotiate some sort of rights to the source code, in case you want or need it later on. Also, ask the developer about his service and support policy. He should agree to support his work (for a reasonable fee) after the implementation is complete. If you use the programmer on a sensitive project, consider having him sign a non-disclosure agreement.

Technical Consultants

From time to time, every company has need for a consultant with special technical expertise. You may need a network engineer to help design and install your LAN, or perhaps a developer with client/server expertise is needed to install your database application server. Whatever the need, hiring a consultant can be a practical way to get over a technical hump in a hurry.

Tips on Hiring a Technical Consultant

Depending on the depth of your project, you may or may not want to use an RFP to solicit proposals from prospective consultants. In general, the more complex the problem, the greater the need for a detailed RFP and proposal. Look closely at the consultant's experience with similar projects. You may even want to verify his references.

The Most Effective Use of Technical Consultants

While you may be tempted to hire a consultant to go complete your project on his own, we recommend you get your own people involved in at least part of the work. In this way, the consultant can share his expertise and train your staff members to do the work themselves in the future. Of course, if the project is a "one-shot deal" where you know you will never need the expertise again, it isn't practical to spend time training your people.

Taking this concept of expertise sharing one step further, we know of one firm that occasionally hires a consultant to act as a technology tutor. The consultant visits the company for a day or two at a time and does a "data dump" on a technical topic of particular interest. This practice is a small price to pay to learn the "how to's" and the promises and pitfalls of technology.

Aside from benefiting from shared expertise, you may want to use a technical consultant to verify your own work. For instance, an outsider can look over your network implementation and strategy and advise you on improvements, enhancements, and future directions. Or, the consultant can analyze your transaction-processing requirements and make recommendations regarding your choice of a back-end database management system. Consultants are often able to look at many issues and alternatives that you may not have considered in making your technology choices.

Special Considerations When Hiring a Technical Consultant

Experience is everything when hiring a technical consultant. You want someone who has a broad knowledge base and yet who has an intimate knowledge of technical details.

If the technical consultant is hired to actually implement some form of technology—such as install a LAN—be sure the person documents his work well to minimize your dependence on the consultant in the future.

Management Consultants

Technology issues often spur management issues. How should the IS department be organized? What positions are required, and how can the employees be adequately compensated? What sort of five-year plan can be developed to prepare for client/server computing? While the issues are certainly technology-related, the fundamental decisions belong to management. You may want to hire a management consultant with a technology background to help sort through these issues.

Tips on Hiring a Management Consultant

Forget the RFP in hiring a management consultant. Instead, find a firm with a good reputation for addressing the issues that you have on your plate. The Big 6 accounting firms are well known in this arena. However, don't exclude the smaller consulting firms, as they may be able to offer you more personal service.

The Most Effective Use of Management Consultants

Management consultants are best used in helping you formulate long-term strategies and deal with "people" issues. They tend to look past the "gee whiz" features of technology and consider the long-range impact that the technology will have on the company's business. They also see people as your most important resource and focus on how to make best use of your work force. This type of consultant is an adviser, not an implementer.

Management consultants are in high demand to help companies through their business reengineering issues. In this case, consultants bring a third-party objectivity that is so important when a company is trying to come up with new solutions to old problems. Outsiders have no ties to how a problem has been addressed in the past and no preconceptions of how it should be addressed in the future.

Special Considerations When Hiring a Management Consultant

Choose a consultant with a good reputation and solid credentials—they will be important in getting executive management to buy into the consultant's recommendations. If the consultant you select is an unknown or has questionable credentials, your executives may be hesitant to support his ideas.

What to Watch for When Using a Consultant

No matter what type of consultant you use, there are a few situations to watch out for, as illustrated below:

- Projects that don't seem to have an end to them, or that seem to grow in scope and magnitude as time goes on
- Projects or tasks that grow in cost beyond reasonable expectations
- Contractors that have an allegiance to a particular vendor or vendors. Your "solution" may start to look exactly like the vendor's suite of products
- Consultants that are limited in their range of solutions
- Consultants that don't have the resources to properly complete your project

- Disputes over poorly defined requirements or expectations
- Costly scope modifications after the project is underway
- Mixed signals from having too many people give directions to the consultant

Resources for Finding an Appropriate Consultant

There is an old joke that says that a consultant is just an out-of-work professional. If that were the case, you could just search the unemployment lines for a consultant. While you might be able to find one there, you want to be sure to find a *good* consultant. There are a number of ways to find just the right consultant for your needs. Here are a few tried and true suggestions for where to look.

- Professional associations: There are a number of professional associations that can recommend members or contacts to meet your needs. A few examples are Data Processing Management Association (DPMA), Microcomputer Managers Association (MMA), Corporate Association of Microcomputer Professionals (CAMP), and Applied Information Management (AIM) Institute. Two organizations that are devoted to consulting are the Professional and Technical Consultants Association and the Independent Computer Consultants Association. There is an organization for just about every discipline within the IT industry.
- User groups: Every large city (along with many smaller ones) has organizations for end users of technology. Consultants, particularly developers, can often be found in the membership rosters.
- Vendors: Your hardware and software vendors can often recommend independent consultants that can address your issues and work on your projects. The vendors may also be able to steer you toward consultants with particular credentials or certifications. For instance, Novell Inc. can probably recommend a Certified NetWare Engineer (CNE) in your area.
- Chambers of Commerce: Many consulting groups register with their local Chamber of Commerce.
- Technical publications: Columnists and writers for technical publications often have or know of consulting businesses. When you read a

trade-journal article that has particular relevance to your own situation, contact the author or editor for suggestions as to consultants.

- "Headhunters" and classified advertisements: These are good sources for contract programmers and technical consultants.

You Don't Have to Go It Alone

Technology and the issues surrounding it can be confusing, especially with the hectic pace of change we face today. Fortunately, there are plenty of people who can help you make sense of it all. Don't be afraid to hire a consultant if you feel you need one. Find the money in the budget—and consider it an investment, rather than an expense.

Before you do hire a consultant, plan carefully how you want to use him, and what objectives you want to meet. You may even want to put your requirements into a formal RFP and accept bids. Look for the consultant with the right skills, knowledge, and experience to complete your project.

Appendix A

A Glossary of Information Technology

Application An executable program that performs useful functions such as manipulating data (e.g., a word-processed document or a spreadsheet). An application is not part of the operating system.

Client A front-end computer for application that handles the presentation of data. Clients usually run on a workstation or desktop computer. See *Server.*

Client/server computing A style of computing where the processing of applications is divided among multiple computers.

Compact disc read-only memory (CD-ROM) An optical disk that holds up to 660 megabytes of data from which information may be read from but to which it may not be written. CD-ROMs are becoming the media of choice to distribute multimedia applications that combine voice, data, and imaging.

Compression Reducing the representation of information but not the information itself. Compression saves transmission time and storage capacity.

CompuServe A commercial online service offering e-mail, conferencing, reference, and other services.

Computer literacy The ability to use desktop computers without fear of making mistakes. Knowledge of the on-and-off switch; of how to use an application; of how to navigate directories and subdirectories; and of how to copy, move, and delete files.

Consolidation Reducing and combining the resources that are required to support an organization's information-processing needs.

Cross-functional team A team consisting of representatives from each functional discipline relevant to a given project. The representatives could come from various professional functions, such as finance, sales, and human resources.

Cyberspace A word coined by William Gibson: a place where e-mail messages, pager signals, fax transmissions, and voice mail are stored.

Database Software designed to catalog and organize large volumes of information such as telephone directories or customer records. A database should be organized to enable you to produce reports and compile their data in ways useful for making corporate decisions.

Demassification Decentralizing and downsizing, or reducing the size of a large corporation's operating units.

Directory A logical structure on a disk drive that is used to group and logically arrange files; it may contain files or other directories.

Document Any information created in an application that you either type, edit, view, or save and that is stored as a file on a disk, such as a letter, spreadsheet, business report, or graph.

Downsizing Moving computer applications to a smaller, less expensive hardware platform, such as a microprocessor-based server.

Electronic bulletin board An online service that allows a caller to exchange messages or to transfer files, among other features.

Electronic checkout scanner A device used in grocery and department stores to scan the universal product code or price tag. It can also enter the price in the register and depletion of inventory (depending on the software used).

Electronic Data Interchange (EDI) Exchanging data and information directly between two computers. The information is generally in a standard data format.

Electronic mail (e-mail) A software application that provides for the direct exchange of messages and files (such as correspondence and word-processed documents) among people and computers.

Empowerment Giving each employee the authority and responsibility to make his or her company more successful and giving him or her the knowledge necessary to make positive decisions.

Function-based organization An organization that aligns representatives within their various professional functions such as finance, marketing, and sales. Also known as a hierarchical organization.

Gigabyte (GB) A gigabyte is 1,024 megabytes or about 1 billion characters of data.

Globalization The trend for technology, manufacturing, and communication to grow available at a global level rather than a local, regional, or national one.

Graphical User Interface (GUI) (pronounced gooey) A system for user interaction with an application that features a choice among menus, icons, and other visual devices rather than a prompt for memorized commands.

Groupware Application software that aids in collaborating on projects.

Hardware The physical equipment that makes up a computer system, such as the keyboard, disk drives, mouse, and monitor.

Imaging Scanning paper documents into electronic pictures for online retrieval and processing. Systems exist that will generate unique patterns for words and images contained in the document in order to allow real-time retrieval in context-sensitive searches.

Insourcing Using internal business and technology resources to accomplish a task or complete a project.

Integrated computer-aided software engineering (I-CASE) A tool that checks specifications against a corporate model in a repository and prepares prototype applications for review.

Joint application design (JAD) A software development technique in which end users and IS personnel jointly develop the specifications for and design of a new application.

Local area network (LAN) A collection of computers and peripheral equipment that share common resources and can communicate with each other. The computers are usually in close proximity to each other.

MCI Mail A commercial e-mail service.

Mainframe A large-scale computer designed to serve the needs of hundreds of users. The processing capability is time-sliced to give all users the appearance of simultaneous computing.

Microcomputer Any small desktop, laptop, hand-held, or palmtop computer using very large scale integration (VLSI) technology for manufacturing its components such as the CPU, disk controller, RAM, and serial and parallel interface electronics.

Middleware products Products that bridge the gap between client/server front ends and back ends and that promote industry standards that developers can write applications to without regard for which front ends and back ends may be in use.

Million instructions per second (MIPS) A standard unit of computing power.

Minicomputer A small centralized-processing computer specialized for specific applications or purposes, such as a centralized word-processing unit.

Mobile computing Exchanging data with computers or people while traveling, away from the corporate structure. This sort of operation is usually accomplished with palmtop computers and, usually, without the aid of telephone lines.

Multimedia The combination of voice, data, or images in an application. It depends on powerful hardware and sophisticated software.

Object-Oriented Programming (OOP) A new concept in which events such as opening a file or clicking a mouse button and the data associated with those events are treated as objects. These objects can be coded as reusable modules.

Outsourcing Farming out all or part of IS operations to a third-party service provider.

Palmtop computer A small portable (about 4 inch by 6 inch) hand-held computer weighing about a pound, which usually runs from a regular AA battery.

Personal computer (PC) A desktop computer with the capacity to be used with many applications such as word processors, spreadsheets, graphics designers, and database management systems.

Ping The process by which a node on a network sends out requests to another node until a response is received, or an interaction between the two nodes in which packets of information are exchanged in response to each other's requests.

Privatization The process by which public enterprises such as governments farm out contracts to private industry through competitive procurements. Privatization relieves government of some of its bureaucracy. It is also known as deregulation.

Process-based organization An organization that takes representatives from each functional discipline and puts them on a team. Also known as a cross-functional organization.

Prodigy An online service offering message-handling and reference services that features a graphical interface.

Productivity paradox The situation in which companies have installed new computer technology but do not see increased user productivity as a result.

RadioMail The company that markets and sells a wireless electronic mail gateway service.

Random access memory (RAM) The internal memory used by a computer for temporary data and program storage.

Rapid application development (RAD) An approach to application development that vastly speeds the overall process.

Reengineering The fundamental rethinking and radical redesign of business processes to achieve dramatic improvements in critical measures of performance, such as cost, quality, service, and speed.

Rightsizing Hosting a computer application on the hardware platform that makes the most sense.

Scanners Any of several devices used to scan written material and convert it to electronic form, such as digital data that can be processed by the computer.

Server The back-end processor of applications that stores and secures the data. See *Client.*

Software The instructions loaded into your computer to make the hardware perform tasks. This would include programs, operating systems (e.g., MS-DOS), device drivers, and applications such as Word for Windows, WordPerfect for Windows, Excel, Lotus 1-2-3, Paradox, Access, and Harvard Graphics, to name a few.

Spreadsheet A software application designed for calculating reports, performing financial analysis, and generating graphs. Some examples of spreadsheet software are Lotus 1-2-3 and Excel.

Terabyte (TB) A unit of 1,024 gigabytes, roughly a trillion characters.

Virtual corporation An organization where the representatives are not necessarily all company employees and do not necessarily work in the same geographic location.

Wide area network (WAN) A collection of computers and peripheral equipment that share common resources and can communicate with each other. The computers may be spread out over a vast geographical area.

Wireless communicating device A radio modem that sends and receives switched-packet radio messages such as e-mail by means of a satellite to and from a designated address.

Word processor An application designed for composing and editing letters, reports, and similar documents. Some examples of word processor software are Microsoft Word for Windows, WordPerfect for Windows, and AmiPro.

Workgroup A collection of people that interact with each other on the job.

Workgroup for Electronic Data Interchange (WEDI) A coalition of government and private organizations organized to encourage EDI adoption in the health care industry.

Workstation A microcomputer attached to a network and used to perform user tasks. See also *Client.*

Appendix B

List of References

"Advertising: Roman Wins WPP Battle." *Wall Street Journal,* November 11, 1992, sec. B, p. 6, col. 3.

"BfG Bank: Gothic Takeovers." *Economist,* vol. 325, iss. 7782, October 24, 1992, pp. 88-89.

"British Airways Said to Consider a Bid for Qantas." *New York Times,* October 28, 1992, sec. D, p. 3, col. 1.

"Computerland's Seven Suggestions for Successful Client/Server Computing." *Computerland Magazine,* January/February 1992.

"Dunhill Is Said to Seek Gucci." *New York Times,* November 2, 1992, sec. D, p. 7, col. 1.

"East German Companies: Private, Perhaps." *Economist,* vol. 325, iss. 7784, November 7, 1992, p. 87.

"Gillette Plan to Buy Parker Pen Receives Scrutiny on Antitrust." *Wall Street Journal,* October 7, 1992, sec. A, p. 11, col. 1.

"Hanson: Hunter's Return." *Economist,* vol. 325, iss. 7780, October 10, 1992, pp. 84-85.

"Kroger Co." *Wall Street Journal,* August 19, 1992, sec. B, p. 4, col. 5.

"Russia: Reform in One City." *Economist,* vol. 325, iss. 7784, November 7, 1992, p. 60.

"World Wire: Argentine Nuclear Privatization." *Wall Street Journal*, October 28, 1992, sec. A, p. 14, col. 3.

"World Wire: Philippines to Speed Sell-Offs." *Wall Street Journal*, November 4, 1992, sec. A., p. 10, col. 4.

"World Wire: Portugal to Privatize Radio." *Wall Street Journal*, October 30, 1992, sec. A, p. 7, col. 5.

Ayre, Rick. "Mail-enabled Applications Help Groups Work Together." *PC Magazine*, October 27, 1992, pp. 268-269.

Bannon, Lisa. "Italy Privatization to Start with Sale of Credito Italiano." *Wall Street Journal*, November 10, 1992, sec. A, p. 20 col. 4.

Belasco, James. *Teaching the Elephant to Dance.* New York: Crown Publishers, 1990.

Berezhnaya, Olga. "All Together to Bankruptcy?" *Moscow News*, Iss. 45, November 8, 1992, p. 11.

Bray, Nicholas. "Makes $1.5 Billion Bid for RHM, Topping Hanson Offer." *Wall Street Journal*, October 30, 1992, sec. A, p. 7, col. 1.

Brousell, David, Elaine Appleton, and Jeff Mood. "Levi Strauss's CIO On: The Technology of Empowerment." *Datamation*, June 1, 1992, vol. 38, no. 12, p. 120 (4). Newton, MA: Dahners Publishing Company.

Bruousell, David R. "Multimedia's Giant Conceptual Leap." *Datamation*, January 15, 1993, p. 114.

Buzzard, James. "The Client/Server Paradigm: Making Sense Out of the Claims." *Data Based Advisor*, August, 1990, p. 72 (8).

Carr, Grace M. "Share and Share Alike." *Lan Technology*, November 1992, pp. 74-81.

Choi, Audrey. "Conflict Besets Eastern German Steel Industry." *Wall Street Journal*, November 4, 1992, sec. A, p. 10, col. 1.

Choi, Audrey. "France to Sell Rhone-Poulenc Shares to Public." *Wall Street Journal*, October 30, 1992, sec. A, p. 9C, col. 3.

Conner, Daryl. *Managing at the Speed of Change.* New York: Random House, 1992.

Davis, Stan, and Bill Davidson. *20-20 Vision: Transform Your Business Today to Succeed in Tomorrow's Economy.* New York: Simon and Schuster, 1991.

Drucker, Peter. "The New Society of Organizations." *Harvard Business Review*, September 1992.

Dunn, Peter. "GenRad Plans 2750 Model to Run Software for 1970." *Electronic News*, vol. 38, iss. 1899, February 17, 1992, p. 18.

Edwards, Billy, and Adrienne Moch. "From Water to Public Works: One City's Privatization Success Story." *American City & County*, vol. 107, iss. 12, November 1992, pp. 38-39.

Edwards, Morris. "Some Notes on Groupware for LANs." *Communications News*, July 1992, p. 42.

Ehrenman, Gayle C. "Meeting Go Electronic, for a Price." *PC Magazine*, May 12, 1992, p. 31.

Eng, Paul M. "Bits & Bytes." *Business Week*, January 25 1993, p. 88D.

Fischer, Bryan J. "Privatization in Poland, Hungary, and Czechoslovakia." *RFE/RL Research Report*, vol. 1, iss. 44, November 6, 1992, pp. 34-39.

Garneau, George. "Zuckerman and the 'Bleeding Dinosaur.'" *Editor & Publisher*, vol. 125, iss. 44, October 31, 1992, pp. 12, 35.

Goff, Leslie. "High School Goes Prime Time." *LAN Magazine*, April 1993, pp. 151-156.

Hammer, Michael. "What Is Reengineering." *Information Week*, May 5, 1992, p. 10.

Herman, James, Christopher Serjak, and Perter Sevcik. "Switched Internets: The Coming Gigabit Revolution in Enterprise Networking." *Distributed Computing Monitor*, February 1993, pp. 3-28.

Hudson, Richard L. "Hanson Offers $1.35 Billion for Bread Baker." *Wall Street Journal*, October 6, 1992, sec. A, p. 15, col. 1.

Job, Mark. "What Is Groupware Anyway?" *Database Advisor*, August 1992, pp. 60-62.

Kantrowitz, Barbara, and Joshua Cooper Ramo. "An Interactive Life." *Newsweek*, May 31, 1993, pp. 42-44.

Kaplan, Alan, Robert Lauriston, and Steve Fox. "Groupware (Buyers Guide)." *PC World*, March 1992, pp. 208-215.

Koehler, Scott. "The Scoop on OOP." *Computerworld*, February 8, 1993, p. 52.

Kramer, Matt. "E-mail May Hold Key to Groupware's Future." *PC Week*, October 26, 1992, p. S20.

Leonard, Jonathan S. "Unions and Employment Growth." *Industry Relations*, vol. 31, iss. 1, Winter 1992, pp. 80-94.

Marshak, Ronni T. "Focusing Our Workgroup Coverage: Concentrating on the 3Cs: Communications, Collaboration, and Coordination." *Patricia Seybold's Office Computing Report,* May 1992, vol. 15, No. 5, pp. 2-3.

Nunamaker, Jay F., Jr., "Automating the Flow: Groupware Goes to Work." *Corporate Computing,* October 1992, pp. 187-191.

Peters, Tom. *Liberation Management.* New York: Random House, 1992.

Popcorn, Faith. *The Popcorn Report.* New York: Doubleday, 1991.

Powell, Bill, Anne Underwood, Seema Nayyar, and Charles Fleming. "Eyes on the Future." *Newsweek,* May 31, 1993, pp. 39-41.

Quinnel, Richard A. "Llama Alert!" *EDN,* June 28, 1990.

Quint, Michael. "Takeover Set for a Bank in Colorado." *New York Times,* November 10, 1992, sec. D, p. 3, col. 3.

Rao, Anand. "Team Spirit." *LAN Magazine,* March 1993, pp. 109-114.

Raskin, Robin. "For the Good of the Group?" *PC Magazine,* January 12, 1993, p. 30.

Senese, Gene. "Get Ready to 'Morph' Yourself." *Computerworld,* June 7, 1993, p. 33.

Senge, Peter. *The Fifth Discipline.* New York: Doubleday, 1990.

Sheridan, John H. "JIT Spells Good Chemistry at Exxon." *Industry Week,* vol. 240, iss. 13, July 1, 1991, pp. 26-28.

Skorstad, Egil. "Mass Production, Flexible Specialization and Just-in-Time: Future Development Trends of Industrial Production and Consequences on Conditions of Work." *Futures,* vol. 23, iss. 10, December 1991, pp. 1075-1084.

Slofstra, Martin. "Is E-mail About to Explode?" *Computing Canada,* February 15, 1993, pp. 1-2.

Spiegel, Leo S. "Expanding Brain Bandwidth." *LAN Technology,* November 1992, vol. 8, no. 12, pp. 39-43.

Sproull, Lee and Sara Keisler. *Connections: The New Ways of Working in the Networked Organization.* Cambridge: MIT Press, 1991.

Stark, David. "Path Dependence and Privatization Strategies in East Central Europe." *East European Politics & Societies,* vol. 6, iss. 1, Winter 1992, pp. 17-54.

Stone, M. David, Steven Chen, and Steve Rigney, "Have Your Computer Call My Computer." *PC Magazine,* February 9, 1993, pp. 271-288.

Stuckey, M. M. *Demassification: A Cost Comparison of Micro vs. Mini.* New York: Fourth Shift, 1989.

Stuckey, M. M. *Demass: Transforming the Dinosaur Corporation.* New York: Productivity Press, 1993.

Toffler, Alvin: *Powershift: Knowledge, Wealth, and Violence at the Edge of the 21st Century.* New York: Bantam Books, 1990.

Valdes, Ray. "Sizing Up Application Frameworks and Class Libraries." *Dr. Dobb's Journal*, vol. 17, iss. 10, October 1992, pp. 18-35.

Weimer, George, Bernie Knill, *et al.* "Integrated Manufacturing: Compressing Time-to-Market Today's Competitive Edge." *Industry Week*, vol. 241, iss. 9, May 4, 1992, pp. IM2-IM16.

Zuboff, Shoshana. *In the Age of the Smart Machine.* New York: Basic Books, 1988.

Index

B

E

M

R

S

U

V

W

To order by phone with Visa or MasterCard, call (916) 786-0426, Monday–Friday, 9 a.m.– 5 p.m. Pacific Standard Time.

To order by mail fill out the information below and send with your remittance to: Prima Publishing, P.O. Box 1260, Rocklin, CA 95677-1260

Quantity	Title	Unit Price	Total
		Subtotal	
		7.25% Sales Tax (CA only)	
		Shipping*	
		Total	

Name

Street Address

City State ZIP

Visa/MC No. Expires

Signature

*$4.00 shipping charge for the first book and $0.50 for each additional book.